T0180605

THE
RISE AND FALL
—— OF THE ——
MEDIA ESTABLISHMENT

THE
RISE AND FALL
OF THE
MEDIA ESTABLISHMENT

Darrell M. West
Brown University

BEDFORD/ST. MARTIN'S   Boston ◆ New York

ISBN 978-1-349-62495-9   ISBN 978-1-137-07405-8 (eBook)
DOI 10.1007/978-1-137-07405-8

For Bedford/St. Martin's

Sponsoring Editor for Political Science: Marilea Polk Fried
Editor: Genevieve Hamilton
Senior Editor, Publishing Services: Douglas Bell
Production Supervisor: Dennis Conroy
Project Management: Publisher's Studio/Stratford Publishing Services, Inc.
Cover Design: Donna Lee Dennison
Cover Photo: Cable News Network Television Studios in Atlanta: Copyright © Franz-Marc
  Frei/CORBIS
Composition: Stratford Publishing Services, Inc.
Printing and Binding: Haddon Craftsman, an RR Donnelley & Sons Company

President: Charles H. Christensen
Editorial Director: Joan E. Feinberg
Director of Marketing: Karen R. Melton
Director of Editing, Design, and Production: Marcia Cohen
Manager, Publishing Services: Emily Berleth

Library of Congress Catalog Card Number: 00-107776

Manufactured in the United States of America.

6 5 4 3 2 1
f e d c b a

For information, write: Bedford/St. Martin's, 75 Arlington Street,
Boston, MA 02116   (617-399-4000)

ISBN: 978-0-312-24777-5   (paperback)

Transferred to Digital Printing 2007

To
www.InsidePolitics.org
for bringing the media revolution
home to me personally.

Preface

Few topics have received more attention than the nature of the American media. Vilified by some and feared by others, the manner in which reporters practice their craft and influence the course of society, culture, and politics has been the subject of many books. Few books, however, bring a historical perspective to the subject, discuss the eras through which journalism has evolved, and rely on a series of case studies to make the argument. This book seeks to fill that void by taking an explicitly historical approach to the question of how reporters do their jobs and what kind of impact they have on politics and society.

American journalists long have been considered a powerful force in the social, cultural, and political life of our nation. With high public esteem, a professional style of coverage, and a monopoly on news gathering, reporters in the twentieth century were major gatekeepers who exercised enormous clout over the framing of national events. But now, another era has dawned in which new media have appeared, the television viewing audience has splintered, citizens have lost confidence in the fairness and objectivity of reporters, coverage is more heterogeneous, and the press is less able to direct the impressions of the general public. Indeed, the very entity popularly known as "the media" seems strangely out-of-date amidst the proliferation of news and entertainment outlets, and the varying degrees of professionalism within the industry.

In this book, I document the rise and fall of the American media establishment from 1789 to 2000. The early days of our country did not feature a powerful media. The partisan excesses of the first few decades of our country's existence and the commercial tabloidism that followed weakened reporters and gave them scant public respect. With their tendency to mimic the views of party leaders and write sensationalistic accounts of crime and vice, nineteenth century journalists simply did not have much credibility or independent power.

In the twentieth century, through the gradual emergence of a professional press, journalists rose as a political force and achieved unprecedented power over the dissemination of the news. Due to changes in the way in which the media were organized, covered the news, and were perceived by citizens, journalists became major power brokers. In the eyes of the general public, few members of society held more respect and were seen as more fair and trustworthy than reporters. Claiming only they could fairly report the news, professional journalists gained great credibility and came to exercise considerable clout over the course of national events.

In the last decade, though, a new stage of media life has developed. With the emergence of hundreds of new media outlets from cable television to talk radio to the Internet, the media establishment has lost control over news reporting and a wide range of media outlets are in a cutthroat competition with one another. In this era, news coverage is less homogenous and professional, and more sensational and tabloid-oriented. Through these changes as well as a decline in public respect for journalists, the power that we have associated with reporters has eroded. The glory days of Edward R. Murrow, Walter Cronkite, and David Broder have given way to Matt Drudge, Rush Limbaugh, and crude jokes by Jay Leno and David Letterman on late-night television.

These developments have profound consequences for our society, culture, and system of government. With the loss of power by the mainstream press, elite journalists no longer are able to dictate how the public responds to national events. The more heterogeneous coverage of the current era has weakened the shared cultural understandings that knit Americans together. Niche-market narrowcasting is in and mass broadcasting is out. The electronic Balkanization which has ensued means citizens are divided by gender, race, lifestyle, and geography in how they see current events. The result is a media system that pleases virtually no one.

In discussing how we reached this state of affairs, I review why the media establishment gained and lost influence over the past two centuries, how these shifts in media power and organization occurred, what the consequences were for those in public life, and what challenges lay ahead for the industry. Using a historical approach based on two dozen comparative case studies from the Founders to current period, I reveal how reporters in different periods covered events of the day and what it meant for the formation of the American state. I conclude the book by noting how the fragmentation of the current media system ironically protects us from many of the problems that worry contemporary observers, such as corporate control over the news and the seeming prevalence of biased, superficial, and unfair coverage. Even though there has been a clear loss of professionalism in the industry as a whole, the excesses of individual outlets are balanced by the dramatic increase in the diversity of outlets, the heterogeneity of coverage, and the declining power and audience share of specific channels. As de Tocqueville concluded of the early American press, there is protection in large numbers in the overall media system.

This book is designed for use in several different courses: introduction to American politics, mass media, campaigns and elections, and American political development, among others. With its emphasis on readable case studies showing how journalists and editors have played different roles at various points in American history, the volume reveals how the media function and why the contemporary media have lost so much respect and clout.

During the course of writing this book, I had help from a number of different people. Alex Slawsby performed yeoman service as my research assistant. He proved quite expert at locating a range of relevant citations and obscure facts for me. Alex also provided a number of helpful suggestions on the general outline of the book.

Although he didn't realize it at the time, Byron Shafer of Nuffield College at Oxford University gave the impetus for this book project by inviting me to co-teach a seminar there during a 1993 sabbatical. One of the lectures I delivered at Oxford outlined the framework for what later became this book.

Over the past few years, students in my campaigns and elections, and mass media courses at Brown University have aided this project by asking compelling questions during discussions of this research project. The overall enthusiasm of these students helped convince me that a book on this subject needed to be written.

James Headley, former political science editor at Bedford/St. Martin's, deserves a thank you for seeing the potential of this book. His successor, Katherine Kurzman, provided detailed comments which made this a better book. The external reviewers made a number of helpful suggestions on how to strengthen the argument of the book: Martin L. Brownstein, Ithaca College; James M. Carlson, Providence College; Jay S. Goodman, Wheaton College; Doris Graber, University of Illinois, Chicago; Matthew J. Lindstrom, Siena College; Robert C. Sahr, Oregon State University; Richard A. Schwarzlose, Northwestern University; Michael C. Tolley, Northeastern University; and John Zaller, UCLA. I am especially grateful to John Zaller, Jim Carlson, Jay Goodman, Doris Graber, and David Jones for their keen insights. Leslie Connor did an excellent job copyediting the manuscript.

Finally, I appreciate the help provided by <www.InsidePolitics.org>, the Department of Political Science, the John Hazen White, Sr. Public Opinion Laboratory, the A. Alfred Taubman Center for Public Policy and American Institutions, and the Undergraduate Research and Teaching Assistantship program at Brown University. I could not have completed this book without the support of these units.

Darrell M. West

Contents

About the Author

Darrell M. West is John Hazen White Professor of Political Science and Public Policy and director of the Taubman Center for Public Policy at Brown University. West received his Ph.D. in political science from Indiana University in 1981, specializing in American politics, mass media, and elections. His current research focuses on the history of the mass media and television advertising in election campaigns.

West is a frequent commentator on media and elections. He has been quoted in the *New York Times, Washington Post,* National Public Radio, and on CNN, among other media outlets. He has served as an election consultant to local television stations in Providence since 1988.

West has written extensively on American politics, mass media, and elections. Previous books include *Air Wars: Television Advertising in Political Campaigns, 1952–1996,* second edition, (1997); *The Sound of Money* (1998), with Burdett Loomis; *Checkbook Democracy: How Money Corrupts Political Campaigns* (2000); and *Patrick Kennedy: The Rise to Power* (2000). He is the developer of <www.InsidePolitics.org>, a Web site that features in-depth information about state and national politics.

THE
RISE AND FALL
—— OF THE ——
MEDIA ESTABLISHMENT

CHAPTER 1

Introduction

On January 17, 1998, a thirty-one-year-old man named Matt Drudge who operated an Internet newsletter out of his Hollywood apartment was about to publish titillating rumors regarding an adulterous sexual relationship between President Bill Clinton and former White House intern Monica Lewinsky. Specializing in hot and often times unconfirmed gossip about celebrities, Drudge loved to beat the mainstream media to the punch. In a new era where anyone with anonymous sources and a computer could reach millions of people, Drudge was becoming a major force. Serving as his own writer, editor, and fact checker, the young man was attracting hundreds of thousands of viewers to his Web site.[1]

Drudge was not the only person on the Clinton trail, though. *Newsweek* reporter Michael Isikoff, a respected newsman who used to write for the *Washington Post*, had spent months interviewing key sources. Other than the president himself, Isikoff knew as much about the Clinton escapade as virtually anyone else in Washington. During the course of his investigations, for example, Isikoff had discovered that Clinton was having an affair with the young intern and that the president privately had been accused of groping the breast of a White House volunteer named Kathleen Willey.

Isikoff, however, faced one disadvantage not shared by Drudge. The *Newsweek* reporter had layers of fact checkers, editors, and lawyers looking over his shoulder to make sure he got the story right. For an article alleging immoral behavior on the part of a sitting president, company executives demanded that every loose end be checked out thoroughly. They did not want to get anything wrong. These safeguards produced an ironic end to the investigation: Fact checking delayed publication just long enough that Drudge was able to scoop *Newsweek* and become the first journalist to broadcast the story of the president's affair to the general public. In so doing, Drudge set off a raucous debate about what was newsworthy and how new technology was transforming the reporting profession.

The press had not always covered political scandals in the same manner or with the same result. Thomas Jefferson, the third president of the United States, was charged with having fathered a child with one of his slaves. And one of our country's most brilliant founders, Alexander Hamilton, was forced to admit to an adulterous relationship with a woman after the lady's husband blackmailed him.[2] In neither case, however, did the leader's public reputation suffer. Both Jefferson and Hamilton maintained high public trust and credibility despite disclosures about

1

their personal lives because press coverage tended to be very partisan during this period and readers exercised due caution about what they read in the paper.

In 1884, Democratic presidential candidate Grover Cleveland was accused of a scandal arising from the fact that in 1874 he had fathered a child out of wedlock with a Buffalo widow named Maria Crofts Halpin. Ten days after he had secured his party's nomination, the *Buffalo Evening Telegraph* published a story entitled "A Terrible Tale. A Dark Chapter in a Public Man's History." Taking aim at Cleveland's pious claims to be a virtuous man intent on reforming the political process, the newspaper detailed what it called "the pitiful story of Maria Halpin and Governor Cleveland's Son."[3]

The late-nineteenth century was an era dominated by press commercialism, strong publishers, the desire for even greater circulations, and sensationalistic coverage. The Buffalo expose was followed by similar stories in other tabloid papers and whispered rumors that the nominee had "consorted with low women." Even though Cleveland accepted financial responsibility for the child after the baby was placed in an orphanage, opponents had a field day taunting him with the catchy ditty "Ma, Ma, Where's My Pa?/Gone to the White House/Ha! Ha! Ha!" Despite the "unprecedented scandal" raised by this event, the reporting did not prevent Cleveland from becoming president.[4] After a bruising campaign filled with negative campaigning on both sides, he beat his opponent, Republican James Blaine, by a razor-thin margin of 23,000 votes to become the first Democratic president in the two decades after the Civil War.

In the early 1960s, President John F. Kennedy engaged in affairs with many different women. There were nude swims at the White House and luncheon trysts with office secretaries.[5] None of these activities, however, were ever made public until years later because the press defined the activities as private behavior not worthy of news coverage. The sixties was the era of the objective press. Professionally trained journalists sought to overcome the tabloid and partisan tendencies of their predecessors. Public lives were sharply delineated from private ones. Sexual behavior was not news unless criminal charges were filed or the actions were too indiscreet.

In 1987, however, the rules on press coverage changed with a vengeance. Gary Hart was the clear front-runner for the Democratic presidential nomination. After his strong primary run in 1984, he was seen as having an excellent chance of becoming president. However, reporters who knew Hart were concerned about his personal character. Rumors had long circulated about the candidate's playboy behavior. According to the Washington grapevine, Hart had engaged in affairs with a long list of women.

At the time, a new press era was emerging, in which journalists were shifting from the goal of objectivity to investigation and interpretation. Seeking evidence of such infidelity, a *Miami Herald* reporter staked out Hart's townhouse. Before long, the writer found the candidate and a young single woman named Donna Rice spending the night together.[6] Confronting the two the next morning as they left the townhouse together, the reporter wrote a story charging Hart with improper behavior. In the eyes of many journalists, Hart's actions were symbolic of a much deeper character flaw based on excessive risk taking and irresponsible personal actions.

Rather than ignore personal conduct, as had been the case throughout much of the era of the objective media, reporters pressed Hart on the accusations of infidelity. "Was he engaged in adultery?" they demanded at a televised press conference. Proclaiming the press had no right to ask such questions, Hart refused to answer. Within a week, though, he was out of the race, utterly unable to get the press to focus on any issue other than his personal fitness for high office.

In the Clinton case, the outpouring of media attention produced the most astonishing resolution of all.[7] Following press revelations about Clinton's affair with Lewinsky, the president's job performance ratings jumped to 67 percent approval and his party actually gained seats in the 1998 election, breaking the normal pattern of midterm losses for the party controlling the presidency. Public support for his presidential performance rose to 72 percent after he was impeached by the House for perjury and obstruction of justice associated with the scandal.[8]

The most surprising development of all was that the loser in this controversy turned out to be the mass media themselves. Rather than blaming Clinton for engaging in bad behavior that led to an avalanche of news articles, ordinary people saw the press as responsible for developments they did not like. In their eyes, reporters were responsible for the feeding frenzy that emerged from the myriad new media outlets. Radio talk show hosts and around-the-clock news stations were examining every minute detail to a nauseating degree. Television and the Internet were putting graphic language about oral sex and masturbation into the public domain. "The media" were fanning the flames of the controversy and getting key details wrong to boot, earning them the public's disdain.

In this book, I examine the rise and fall of the news media establishment as a major political power from 1789 to 2000. After looking at two centuries of reporting on American politicians from George Washington to Bill Clinton, I argue there have been systematic differences across historical eras in how reporters covered events and the amount of influence that the media exercised. In the nineteenth century, for instance, reporters were not very powerful. Owing to their origins as party sycophants and practitioners of commercial tabloidism, journalists earned limited respect and held little independent clout with publishers or the general public. Hence, when personal charges were made about such respected men as Jefferson, Hamilton, and Cleveland, the coverage was taken with a grain of salt and did little to affect public reputations.

It was not until the twentieth century, when reporters altered their style of reporting and gained public confidence, that they emerged as an independent voice in public life. These were the glory days of American journalists, when news reporters were viewed as professionals and well-respected by the public. Because of their near-monopoly over news gathering, journalists became major power brokers and shaped the country's agenda and political life.

Now in the twenty-first century, due to structural changes in the media and a style of coverage despised by the public, the mainstream press have fallen as a political power. Fragmented by the development of new media outlets, such as cable, talk radio, and the Internet, their coverage is much less professional and homogeneous. Anyone can report the news, not just trained journalists employed by professional news organizations. In this era, reporting of presidential scandals does not produce the same impact because citizens no longer trust the media messenger. We have gone from a period where the media exercised great influence to one in which reporters have lost the public credibility that gave them distinctive clout. My aim in this book is to show how these stages evolved, what these historical developments mean for the contemporary media, and how these changes affect our society, culture, and system of government.

To document my argument, I adopt a historical approach using comparative case studies of press coverage of two dozen controversies. For example, to see how journalists functioned during our country's early development, I look at the bitter rivalry between Alexander Hamilton and Thomas Jefferson which marked our nation's formation, the Alien and Sedition Laws, the Jeffersonian era, and the Jacksonian Revolution. Later in the nineteenth century, I focus on the rise of technology, the role of powerful independent editors such as James Gordon Bennett, Horace Greeley, Henry Jarvis Raymond, Joseph Pulitzer, and William Randolph Hearst, and case studies of media coverage of slavery, the Civil War, Reconstruction, and the Spanish-American War. In the twentieth century, I study reporting about the Progressive movement, World War II, Vietnam, Watergate, Hart's campaign, William Kennedy Smith's rape trial, Clarence Thomas's Supreme Court confirmation hearings, O.J. Simpson's murder trial, the Clinton-Lewinsky scandal, and John F. Kennedy, Jr.'s plane crash and death.

My goal in assessing coverage of these cases as well as others is to see how media outlets functioned, how the public responded, and what press reporting tells us about the media of each time period. By looking at media coverage and organizations over a period of two centuries, we can study long-term changes in the media and understand why things evolved the way they did. Each era of news gathering left an indelible mark on the political system of that time. In each epoch, there were distinctive patterns of news gathering, different definitions of the news, shifting dilemmas for government officials, and varying reactions from the public in terms of how the media were viewed. Change in media systems occurred through an elaborate interplay of new technologies (such as the telegraph in 1844, the telephone in 1876, the expansion of the railroads, the rise of radio in the 1920s, coast-to-coast television broadcasting in 1946, and the World Wide Web in 1991), the ability of individual politicians, publishers, editors, and reporters to transform media organizations, and the existence of broader political, social, and economic forces such as education, professionalism, war, and market competition that shaped the environment in which media organizations functioned.

I rely on a number of different kinds of materials to look at how particular media systems performed. For example, I study press memoirs to get a sense of what motivated major editors and reporters. I look at histories of press systems and controversial events for contemporaneous accounts of the role of the media. I use data on the diffusion of technological innovations and the impact of changing education levels, professionalism, war, and market competition to see how each affected news coverage. I include information from polls, Nielsen ratings, memoirs, and historical accounts to gauge public reactions to the press of each era. I also draw on original newspaper, television, and Internet reports to demonstrate changing styles of coverage.

In reviewing the historical evolution of the media, I argue that there have been five general stages: the partisan media from the 1790s to the 1840s, the commercial media from the 1840s to the 1920s, the objective media from the 1920s to the 1970s, the interpretive media from the 1970s through the 1980s, and the fragmented media which arose in the 1990s. In offering general time periods for these stages, I do not mean to suggest clear delineations of media stages that are analogous to eras of party realignments. Media stages unfold more chaotically and more gradually than do party periods marked by major elections and cataclysmic events such as the Civil War and the Great Depression. Media epochs are not well-defined and generally do not have clear start and end-dates. Such periods typically represent a mishmash of different kinds of coverage. For example, one can find elements of tabloid coverage during the era of the objective media, as well as partisanship during the contemporary fragmented media. Yet despite these anomalies, in looking at the media's historical development, there are discernible patterns that help us to think about the long-term transformation of the media and the underlying dynamics that resulted from these changes.

The book is organized around these five historical stages, complete with relevant case studies and data from each period. Chapter 2 examines the first historical stage known as the *partisan media*. In this period covering the first few decades of our country's existence, I argue that reporters held little independent influence. Major press outlets were sponsored by political parties and were seen as vehicles to attract voter support to that party. Newspapers had clear partisan viewpoints and served as vehicles for the fulfillment of partisan objectives. Political coverage attempted to mobilize supporters of the party, and journalists served as party sycophants. Coverage of major controversies such as the Hamilton-Jefferson rivalry reflected that kind of perspective.

From the 1840s to the 1920s, the press evolved into a slightly more powerful stage known as the *commercial media* (see Chapter 3). With the extension of the penny press, the development of the telegraph in 1844, the expansion of the railroad, and the industrialization of America, the media became more commercially directed and tabloid-oriented. This period saw our country's first wire service, the Associated Press, which helped broaden the audience for news. Several national dailies became prominent during this period, which aided the process of nation-building that the country was undergoing. To illustrate these changes, I look at the impact of technology on news transmission, the rise of great publishers and editors, and formative events such as slavery, the Civil War, and Reconstruction.

In Chapter 4, I review the period from the 1920s to the 1970s, when the press evolved into a powerful entity known as the *objective media*. As in occupations such as law, medicine, business, and education, during this period we see the professionalization of the industry. Journalism schools arose to train reporters in professional standards, such as fairness, objectivity, and the effort to avoid direct partisan biases. News departments were separated from editorial departments. Newspapers were free to editorialize on the editorial page, but not on the front page. The goal of reporters was to search for the truth and evaluate government performance along clear and objective standards. While no one concluded that reporters actually achieved these lofty ambitions, the effort at objectivity and the general professionalism of the coverage enhanced the credibility of journalists and gave them unusual respect among the American public.

Chapter 5 looks at the *interpretive media*, which arose during the 1970s and 1980s. This period featured still another stage that was fundamentally different from earlier eras of American politics. Following intellectual trends in many academic fields, which placed value on contextualizing particular events, reporters became more interpretative in their approach. Rather than taking candidate and leadership statements at face value, journalists began to write news analysis pieces and "Ad Watches" designed to put political activities in a larger context. Pundit analysis and interpretation became more common, as reporters sought to bring their personal knowledge of and familiarity with news makers to the attention of readers and viewers in order to inform them about news activities. With the goals of analyzing and interpreting, however, came a more subjective style of coverage that annoyed the public. Seeing this style of reporting as being less factual and more intrusive into personal backgrounds, readers and viewers started to develop more negative views about the fairness and balance of American media coverage, which undermined the ability of reporters to affect people's impressions.

The most recent stage of American history, which developed in the 1990s, I describe as the *fragmented media* (see Chapter 6). In this period of declining media power, we see the emergence of new television networks, cable outlets, satellite technologies, and the World Wide Web. Along with an expansion of local news services, these new viewing options have dramatically increased the number of news channels, changed the relationship between the media core and periphery, weakened the overall professionalism of news presentations, and undermined media influence. As illustrated by the infamous Drudge Report, the days when ABC, CBS, NBC, the *New York Times*, and the *Washington Post* could dominate news gathering have given way to tabloid journalism, cutthroat media competition, and an era when obscure press outlets can break major news stories. The mainstream media establishment has lost its control over the news as a wide variety of new actors have arisen to report the events of the day. Coverage in this new media environment is less professional, more heterogeneous, and more tabloid-oriented. Parts of it also are more partisan, thereby echoing elements of our country's first media system.

In Chapter 7 I discuss the future of the American media. Given the fragmentation of the media industry and the sharp decline in public respect, journalists have lost considerable influence over the political process. The ability of reporters to shape public impressions has dropped substantially due to a decline in the homogeneity of coverage, the drop in media professionalism, the rise of tabloid journalism, the loss of public trust, confidence, and respect for the media business as a whole, and the simple fact that, today, anyone with a Web site can report the news.

In the next few years, the most important challenge facing the media will be a backlash from dissatisfied viewers and public officials. There is a risk of tougher libel laws, public boycotts arising from the prevalence of small niche markets, and the potential end of a shared sense of American culture due to the large number of heterogeneous news outlets. I discuss future scenarios facing the media, such as the continuation of cutthroat competition, an industry re-concentration around media moguls Eisner, Murdoch, Turner, and Gates, and a possible rise in the United States of a European-style partisan press. Whatever way the media develops, the communications industry will have consequences for how the political system functions.

NOTES

1. See Francis Clines, "Gossip Guru Stars in Two Roles at Courthouse," *New York Times*, March 12, 1998, A25; George Lardner Jr., "The Presidential Scandal's Producer and Publicist," *Washington Post*, November 17, 1998, A1; Marvin Kalb, "The Rise of the 'New News': A Case Study of Two Root Causes of the Modern Scandal Coverage," Discussion Paper D-34, Joan Shorenstein Center on Press, Politics, and Public Policy, October 1998; and Bill Kovach and Tom Rosenstiel, *Warp Speed*, (New York: Century Foundation Press, 1999).
2. Dinitia Smith and Nicholas Wade, "DNA Test Finds Evidence of Jefferson Child by Slave," *New York Times*, November 1, 1998, A1; Robert Pear, "Founding Fathers Are Used to Build a Case for Clinton," *New York Times*, October 4, 1998, 37; and Richard Brookhiser, "Alexander Hamilton's Original Sin," *George*, April 1998, 85–87.
3. "A Terrible Tale. A Dark Chapter in a Public Man's History," *Buffalo Evening Telegraph*, July 21, 1884, 1; and John Tebbel and Sarah Miles Watts, *The Press and the Presidency: From George Washington to Ronald Reagan* (New York: Oxford University Press, 1985), 257, 258, 262.
4. Dave Florek, "In Buffalo's Own Scandal about Presidential Sex, Grover Got the Spin Just Right," *Buffalo News*, May 13, 1998, 2B. Also see James Pollard, *The Presidents and the Press* (New York: Macmillan, 1947), 500–501.
5. Seymour Hersh, *The Dark Side of Camelot* (New York: Little, Brown, 1997). For a view of the current generation of Kennedys, see Darrell M. West, *Patrick Kennedy: The Rise to Power* (Englewood Cliffs, N.J.: Prentice Hall, 2000).
6. Jim McGee, Tom Fiedler, and James Savage, "The Gary Hart Story: How It Happened," *Miami Herald*, May 10, 1987, 1; and Robin Toner, "Hart Drops Race for White House in a Defiant Mood," *New York Times*, May 9, 1987, 1. For a retrospective interview with Hart a decade after the fact, see John Kennedy interview, "The Hart of the Matter," *George*, April 1998, 98–104, 139.
7. "A President Deep in Trouble; The Allegation That He Had an Affair with a White House Intern — and Then Urged Her to Lie — Swept Washington and the Nation," *Minneapolis Star Tribune*, January 22, 1998, 1. Also see Francis Clines, "Testing of a President: The Accusers, Jones' Lawyers Issue Files Alleging Clinton Pattern of Harassment of Women," *New York Times*, March 14, 1998, A1; John Henry, "Willey Accuses Clinton of Lying about Incident; Former White House Volunteer Claims the President Groped Her," *Houston Chronicle*, March 16, 1998, A1; and Tim Weiner with Neil Lewis, "How Legal Paths of Jones and Lewinsky Joined," *New York Times*, February 9, 1998, A14. Also see David Brock, "Living with the Clintons: Bill's Arkansas Bodyguards Tell the Story the Press Missed," *American Spectator*, January 1994, 18–30.
8. Richard Morin and Claudia Deane, "President's Popularity Hits New Highs," *Washington Post*, February 1, 1998, A1; Adam Nagourney and Michael Kagay, "Public Support for the President, and for Closure, Emerges Unshaken," *New York Times*, December 21, 1998, A21; and Regina Lawrence, Lance Bennett, and Valerie Hunt, "Making Sense of Monica: Media Politics and the Lewinsky Scandal" (paper presented at the 1999 annual meeting of the American Political Science Association, Atlanta, September 2–5).

CHAPTER 2

The Partisan Media

Shortly into George Washington's presidency, a fundamental debate over the proper role of the federal government divided political leaders of the day. One faction led by Thomas Jefferson was dubious about the need for a strong central authority, which smacked too much of the monarchy Americans had just bested in the Revolutionary War. The Jeffersonians were agrarians, who preferred a decentralized government based on grassroots authority, local banks, and state militias. They were suspicious of concentrated power in any form.

Another group directed by Alexander Hamilton, however, felt the new country needed a strong core to overcome the weak government established in 1781 right after the Revolutionary War. The framework ratified under the Articles of Confederation had proven far too ineffective at dealing with relations among the states. If the country was to grow and prosper, and overcome the petty bickering of sectional interests, a strong national authority was necessary. There should be a standing army, a navy, and a national bank, in order to push American interests. The central government should be given clear powers.

These emerging political factions were noteworthy because the country's Founders had gone to great lengths after the rebellion against England to warn against partisan groupings. James Madison had complained sternly about "factions" in *The Federalist Papers*, the set of newspaper articles written to rally support for the new Constitution.[1] President Washington himself was concerned about personal and ideological divisions in the nation's political community.

It was no accident that the word *party* did not appear in the United States Constitution. The men who would govern, the Founders believed, should be leaders of strong reputations, well-known and well-respected in their communities. The president should be a "great man" who would lead his peers, not a party politician who gained advantage through intrigue and maneuvering.

In the first American election, the system operated pretty much as planned. Washington was chosen president with little disagreement. With a sparkling war record and clear leadership skills, no one seriously opposed his selection as chief executive. He was exactly the type of president the Founders had envisioned.

But there was little tranquility within his cabinet. Almost immediately, fighting between Jefferson, the secretary of state, and Hamilton, the secretary of the treasury, arose over a range of issues. There were fights over what to do about excise taxes and relations with England and France. As national treasurer Hamilton had created controversy in 1794 when the army broke up the Whiskey Rebellion, a group of Pennsylvania residents opposed to his 25 percent tax on spirits. Jefferson did not support the excise tax and bitterly opposed the use of federal troops against local civilians. Not only was it tyrannical, he felt, it was a usurpation of the autonomy of local militias, who should handle these types of problems.[2]

The two men also bickered over foreign policy. Hamilton's perspective was much more sympathetic to aristocratic England, while Jefferson was an open admirer of France, which had helped America win the Revolutionary War, and was undergoing its own democratic revolution. Having spent considerable time in Paris, Jefferson appreciated that country's move away from monarchy in the 1790s. Although troubled by the violence of the French Revolution, he counseled American leaders to seek closer relations with France rather than royalist England, as Hamilton preferred.

Over the course of several years, these battles produced different political groupings and led to the creation of large numbers of newspapers around the country that reflected partisan points of view. Most large cities had up to half a dozen competing newspapers, each of which presented a different perspective to its readers. The vigorous competition and partisan reporting of the early American press became hallmarks of the first few decades of the new republic. Bitter, personal attacks from opinionated newspaper editors were routine and marked the political discourse of the day.

As demonstrated in this chapter, the goal of many journalists of this period was direct advocacy and support for the positions taken by the political party with which they were affiliated. There was little effort at objectivity on the part of reporters, which limited their overall credibility with readers. Indeed, many newspapers were created and later subsidized through the direct financial sponsorship of political patrons. This type of communications system made it virtually impossible for the press to exercise much political power. Unlike later periods of American life when journalists held tremendous power, neither reporters, nor editors, nor publishers displayed any independent judgment. They were controlled by forces from outside the industry, namely politicians, government officials, and party organizations. Combined with low subscription levels, the large number of newspapers, the heterogeneity of coverage, and dependence on partisan financial sponsors, the nascent press was troublesome to individual politicians, but not particularly powerful in shaping public impressions as a whole.

DUELING NEWSPAPERS: HAMILTON VERSUS JEFFERSON

No rivalry was more central to the early fortunes of the new government than that between Hamilton and Jefferson. The two men were strong-willed, passionate about politics, and eager to leave their mark on the system. Both leaders cared deeply about the country, but they had very different ideas about how the nation should develop. Hamilton did not trust ordinary citizens and wanted a government dominated by merchants and commercial interests; men of good breeding should run state affairs, he felt. Early in constitutional deliberations, he had expressed support for a lifetime term for American presidents, as opposed to the four-year term that eventually was adopted.

Jefferson, in contrast, was a democrat who mistrusted centralized authority. He did not approve of Hamilton's monarchical tendencies and his adversary's efforts to strengthen the powers of the chief executive. Jefferson was an agrarian who wanted a limited national government. He opposed a national bank and a standing national army and navy. While Hamilton's power base was in the cities, Jefferson's was centered in rural, farming areas, where most of the population lived. There was a stark personal rivalry between the two officials as well. Hamilton was dictatorial and opinionated, while Jefferson displayed far more skill in dealing with other individuals and in assessing the national political climate. Both Hamilton and Jefferson were ambitious men who saw themselves as prominent leading figures. Each eyed the other as an opponent for future national influence.

In the context of this personal and political rivalry, it was little surprise that each would seek to rally support for their respective positions through the major media of the day. Newspapers were not foreign to Hamilton. In fact, he was the first national leader to see the value of using a newspaper as a mouthpiece. In 1787 and 1788, Hamilton, John Jay, and James Madison had helped publicize the cause by penning a series of anonymous essays extolling constitutional ratification in *The Independent Journal*, a New York City newspaper. These essays later were collected together and reprinted in book form as the classic publication, *The Federalist Papers*.[3] When Washington became president, he chose the brilliant but acerbic and opinionated Hamilton as his secretary of the treasury. From that position in government, Hamilton continued his use of newspapers as a promotional tool.

On April 15, 1789, shortly before Washington took office, *The Gazette of the United States* was established in New York City, then the seat of the national government. Its prospectus clearly stated its objective "to be the organ of the government," printing debates and important papers, and writing thoughtful articles on government.[4]

At the rather costly price of six cents per copy, which was well beyond the reach of ordinary citizens, the paper's audience consisted of people of wealth and culture. This was well before the emergence of mass circulation newspapers. It would not be until the time of Andrew Jackson when newspapers could be bought for a penny that they became accessible to more ordinary people.[5] Just 2 percent of the populace in 1788 — roughly 77,000 people out of the overall population of 3,660,000 — subscribed to a paper (see Appendix Figure A.1).[6] These low subscription levels obviously limited the ability of journalists to reach a mass audience and weakened the ability of the press to be a powerful political agent.

Edited by John Fenno, a school teacher from Boston, *The Gazette* was the voice of the Federalist Party. Its task was to support Hamilton's philosophy of a strong central government guided by the aristocracy. To ensure that Fenno understood his partisan responsibilities, Hamilton kept Fenno on the federal payroll as "the printer" to the Treasury Department at an annual salary of $2,500.[7] When the seat of government moved to Philadelphia, *The Gazette* followed, appearing with a Philadelphia address on April 14, 1790.

Jefferson soon grasped the significance of this publication. Calling *The Gazette* "a paper of pure Toryism, disseminating the doctrine of monarchy, aristocracy, and exclusion of the people," Jefferson moved to establish his own press organ. Jefferson returned from a visit to France in 1790, then in the midst of the French Revolution, amazed at the American national dialogue. Attending a number of dinner parties in Washington with prominent political officials, Jefferson said, "I cannot describe the wonder and mortification with which the table conversation filled me. Politics were the chief topic and a preference for kingly over republican government was evidently the favorite sentiment."[8]

The next year, on October 31, 1791, Philip Freneau, who held a $250-a-year clerkship for foreign languages in the State Department, established a rival newspaper, *The National Gazette*, with the active support of Jefferson. Its goal, according to Jefferson, was to be a "Whig vehicle of intelligence."[9] Freneau had been a Princeton classmate of James Madison's, who had recommended him to Jefferson. One of *The National Gazette*'s early issues proclaimed its philosophy that "public opinion sets the bounds to every government, and is the real sovereign of every free one."[10]

Jefferson long had loathed what he saw as Hamilton's monarchical tendencies. During constitutional deliberations, Jefferson had insisted on amendments protecting freedom of speech and freedom of the press in part because of his fears regarding these points. His correspondence from abroad reveals the importance he attached to newspapers as a public vehicle. Writing from Paris on October 13, 1785, about altering public sentiments, Jefferson declared that "the most effectual engines for this purpose are the public papers." Referring to the British nation, he said, "you know well that that government always kept a kind of standing army of newswriters, who, without any regard to truth, or to what should be like truth, invented and put into the papers whatever might serve the ministers."[11] It therefore was no surprise a few years later that he established his own newspaper in an attempt to influence the country's political dialogue.

Hamilton's and Jefferson's dueling papers covered the news of the day in purely partisan terms. The *Gazette* referred to Jefferson's friend Freneau as a "fauning parasite" and one of the "mad dogs" because of his attachment to Jefferson. Freneau replied by accusing his rival editor of "preach[ing] up in favor of monarchs and titles."[12] During this period, there was no distinction between news coverage and editorial opinion. Commentary was interspersed throughout news stories of the day because partisan advocacy was the clear goal. Fenno proselytized in favor of a monarchy, writing, "a king at the head of the nation to whom all men of property cling with the consciousness that all property will be set afloat with the government, is able to crush the first rising against the laws."[13] Without the benefit of mass circulations or much in the way of commercial advertising, editors were dependent on government printing contracts and political patronage.

Bitter condemnation of men in public life was the major tool by which editors and publishers sought to belittle the ideas of the opposition. Leaders from Washington to Jefferson to Hamilton were vilified in boldly personal attacks. Freneau eagerly castigated Washington for everything from "overdrawing his salary" to supporting "unconstitutional" acts.[14] Freneau even went so far as to write a satirical poem complaining about Washington's monarchical tendencies. Entitled "To a

Would-be Great Man," the sonnet read "When you tell us of kings,/And such petty things, Good Mercy!/ how brilliant your pages!"[15]

Even Fenno, one of the most visible practitioners of the craft, complained about press coverage. Writing near the turn of the century, he conceded that, "The American newspapers are the most base, false, servile and venal publications that ever polluted the fountains of society."[16] By 1795, abusive personal attacks on all the leading political figures of the day were more the rule than the exception. This partisan rhetoric was read with a jaundiced eye by many. Knowing that each outlet was financially supported by major party outlets and committed to extolling that party's point of view, voters often read more than one paper to get a balanced view. Few took the vitriolic attacks as anything other than party sniping from paid spokespeople.

Not only were there commentaries on the personalities of the day, the two editors and other papers affiliated with them jockeyed over government policy. Fenno supported the Hamiltonian philosophy of national control modeled after England. Once, he wrote "take away thrones and crowns from among men and there will soon be an end of all dominion and justice."[17] Freneau favored Jeffersonian principles of popular control and respect for ordinary people. He bitterly opposed Hamilton's proposal for a Bank of the United States. In a letter to a Paris friend, Jefferson complained that "the Tory paper, Fenno's, rarely admits any thing which defends the present form of government." Instead, Jefferson wrote, "They pant after union with England."[18]

The fight between the two leaders and their respective papers became so bitter that President Washington was forced to call in his two cabinet secretaries and ask them to cease and desist their partisan quarrels. Seeing tremendous personal and policy stakes in the discussion, each of course refused. Jefferson strongly defended Freneau on the grounds that "his paper has saved our Constitution, which was galloping fast into monarchy."[19]

But Freneau's paper, *The National Gazette*, did not last long, folding on October 26, 1793. The outbreak of yellow fever in Philadelphia was a major contributing factor. Many worried that the newspapers of the day were transmitting the contagion throughout the city. Yet Freneau's crisis was not just health-related. When Jefferson resigned that year as secretary of state, it had the immediate effect of removing Freneau from the government payroll and ending his government subsidy. Fenno's paper continued until the editor's death in 1798, when the paper was taken over by his son, John Ward Fenno.[20]

Despite its demise, Freneau's paper had an effect on his political era. According to Frank Luther Mott, one of the leading historians of that period, "His paper widened the breach between Hamilton and Jefferson, [and] was influential in consolidating the Republican party as an effective opposition."[21] As the next section shows, Republican forces helped end the then-dominant political power of the Federalist Party.

THE ALIEN AND SEDITION LAWS

In the aftermath of Washington's political retirement came a ferocious succession battle that John Adams, vice president to Washington, won by three Electoral College votes over Jefferson. Elected in 1796 with support from nearly 80 percent of the nation's newspapers, Adams was a close ally of Hamilton and a skeptic of close ties to France, which many Jeffersonians avidly sought.

France's bloody republican revolution scared Adams and many other American leaders and convinced them they had to fight hard for Federalist principles. Then a French scandal gave Adams a major opportunity for political gain. Secret dispatches revealed that French foreign minister Charles-Maurice Talleyrand was attempting to extort money from the American ambassador to Paris. In the scandal known as the XYZ affair, so named for the three agents who approached the American ambassador, the French diplomat vowed to receive our country's representative only if paid a personal bribe. Adams jumped on this effrontery to national pride to gain popular support for the Federalists in 1798 and turn the public against Jefferson's Republican party, which was seen as sympathetic to the French government.[22] The president broke the treaties with France and moved to raise a permanent army and navy to defend national interests.

Sensing the possibility of a great political triumph, Adams embarked on a step that later would cost him his presidency and lead to the long-term decline of the Federalist Party. Taking advantage of short-term public dissatisfaction with the French and their republican allies in the press, Adams proposed legislation that would deal with the French ideas of liberty and free speech. In the president's mind, press freedom of expression had gotten way out of hand.

Despite the fact that his personal popularity was at an all-time high, vile and personal condemnations of Adams were routine in many of the country's newspapers.[23] According to Adams and his friends, partisan editors were ruining the national dialogue and disrupting the country's unity. And there was a national security justification; war with France was a distinct possibility.[24]

For these reasons, the Federalist president proposed several major pieces of legislation to save America from "the evils of unlimited democracy."[25] The Alien Act of 1798 allowed the president to order out of the country any alien deemed "dangerous to the peace and safety of the United States." There were around 25,000 aliens living in the United States at the time. About 10 percent of the country's two hundred newspaper editors were not U.S. citizens, and most of them leaned Jeffersonian in their political philosophy.[26] Meanwhile, the Sedition Act of 1798 made it a crime to conspire against the government and "write, print, publish or quote any false scandal or scurrilous writings against the government of the United States, the President or either House of Congress."[27] Together, these acts represented the boldest assault on freedom of expression the young country had yet experienced.

Designed by Federalist sympathizers, who controlled Congress and the presidency, the Alien and Sedition Acts were designed to curtail the increasingly sharp, partisan comments that were flowing from republican editors and opinion leaders.

With the retirement of Freneau, two editors — first Benjamin Bache, the grandson of Benjamin Franklin, and upon his death from the yellow fever in 1798, William Duane — were the leading republican critics of the Federalist government. Using a newspaper called the *Philadelphia General Advertiser* (popularly known as *Aurora*), which started in 1794, these editors bitterly castigated Hamilton, Adams, and the Federalist Party in general.[28]

Bache even had the temerity to go after the national icon, Washington. Hearing of the president's intention to retire from public life in 1796, Bache wrote, "If ever a nation was deceived by a man, the American nation has been deceived by Washington. Let his conduct, then, be an example to future ages; let it serve to be a warning that no man may be an idol."[29] In 1798, after Bache published a secret letter from Talleyrand urging a reconciliation between France and the United States, the government charged him with treason. But the editor died shortly thereafter.

He was succeeded at the paper by Duane, who had been born in New York and served for a time as parliamentary reporter in London. Disgusted by ill treatment while a reporter in India, he returned to the United States in 1796, where he started writing for the *Aurora*. Like Bache, Duane was a sharp and persistent critic of the Adams administration who regularly skewered the Federalist president.[30] Following passage of the Alien and Sedition Acts, Timothy Pickering, Adams' secretary of state, argued that Duane proclaimed to be an American citizen but that he had been educated in Ireland, lived in England until after the Revolutionary War, and had worked in India before returning to the United States. Writing to the president, Pickering concluded that Duane actually was "a British subject, and, as an alien, liable to be banished from the United States." In 1799, Adams wrote back that "the matchless effrontery of this Duane merits the execution of the Alien Law. I am very willing to try its strength on him."[31]

Typical of an era characterized by sharp partisanship, the Alien and Sedition Acts were enforced mainly against republican editors. For example, Charles Holt, editor of the *New London (Conn.) Bee*, received three months in jail and a fine of $200 for censuring the president and urging men not to enlist in the army.[32] Using their control of the legal process, local magistrates in some communities selected only Federalists' loyalists for juries, and tried and convicted ten editors and printers, all of whom were republican. A number of others were tried, but not convicted.[33]

The pure partisanship of the prosecutions resulted in a tremendous backlash against Adams personally and the Federalist Party in general. Jefferson used public worry over the blatant partisan abuse of the legal establishment and more general concerns about freedom of expression to triumph. In the election of 1800, Jefferson was elected president and pardoned all those convicted under what he called an "unauthorized act of Congress."[34] The new Congress then repealed the unpopular legislation and ended the dismal government effort to crack down on press freedom.

THE JEFFERSONIAN ERA

Jefferson's ascension to the presidency at the turn of the century put in office a man committed to a free press. Writing to Elbridge Gerry shortly after his inauguration,

Jefferson proclaimed, "The right of opinion should suffer no invasion from me."[35] But this sentiment soon would be sorely tested.

The 1800 election did not stop the bitter partisanship that characterized press coverage during this era. Federalist papers subjected Jefferson to the same torrent of scurrilous rumors that the republican press had turned on Washington and Adams. During the 1800 campaign, a Federalist paper wrote that "should the Infidel Jefferson be elected to the Presidency, the seal of death is that moment set on our holy religion."[36]

The coverage so upset Jefferson that he once suggested papers should be divided into four chapters with the headings of truths, probabilities, possibilities, and lies. The first chapter, he joked, "would be very short."[37] Yet the blistering newspaper coverage was no joking matter to the new leader.

The president's longtime foe Hamilton raised money to start a new paper in 1801 called the *New York Evening Post*. Following the Federalist election debacle of 1800, Hamilton felt his party needed a press organ more than ever. Each evening, Hamilton summoned his editor, William Coleman, so that Hamilton could dictate the next day's editorials.[38] These editorials then were picked up for publication by other Federalist papers around the country. It was estimated that despite the election outcome against Adams, 60 percent of the country's papers continued to support the Federalist Party.[39]

In 1802, after Jefferson had passed him over for the Richmond, Virginia, postmastership, journalist James T. Callender began to write mild criticisms of Jefferson's administration. Republican papers responded by attacking Callender. The Scottish-born newspaperman had given his wife syphilis, they noted, and left her to die on a "loathsome bed" while their children starved.[40]

In response, a few days later, the enraged Callender published a story alleging that Jefferson had fathered several children with one of his 130 slaves, Sally Hemings (a charge that two centuries later was corroborated through DNA testing of the couple's distant descendants). Hamilton's paper, the *New York Evening Post*, reprinted the story and gave it wide circulation, as did overseas newspapers. In addition to his personal vendetta against Jefferson, Callender was an avowed racist who detested miscegenation. In print, he referred to Hemings as "Dusky Sally," "Black Sal," and "a slut as common as the pavement."[41] The scandalous press charges did little to hurt Jefferson's popularity, though. Befitting an era of a weak press that was held in low esteem by the general public, the negative newspaper coverage did not prevent Jefferson's landslide reelection two years later.

The loathsome media accusations, however, upset Jefferson. Right before the 1804 election, he wrote a letter to Mrs. John Adams claiming states had the right to restrict press freedom. In a September 11, 1804, correspondence, he argued that though Congress could not abridge freedom of the press, states could do so under certain circumstances: "They have accordingly, all of them, made provisions for punishing slander, which those who have time and inclination, resort to for the vindication of their characters."[42]

Echoing these sentiments, some of Jefferson's allies decided to make an example of a Federalist journalist who in their eyes had gone too far in criticizing Jefferson. When Federalist editor Harry Croswell of the *Hudson (N.Y.) Balance* printed that Jefferson had paid the notorious editor of the *Richmond Examiner* to call Washington "a traitor, a robber, a perjurer" and Adams "a hoary-headed incendiary," Croswell was sued for libel in the 1804 case of *People v. Croswell* and found guilty.[43] In his appeal, Croswell was represented by Hamilton, perhaps one of the most ironic events of the country's young history, but the appeals court upheld the verdict.

At this time, most major cities had intense newspaper competition owing to the existence of multiple papers in each locale. For example, Philadelphia had six daily papers, while New York had five, Baltimore three, and Charleston two.[44] With the 1803 Louisiana Purchase from France and the westward expansion that ensued, newspapers arose throughout the new territory in large numbers. Indeed, as people moved west, the desire for news from home increased rapidly. But most of these newspapers maintained an explicit partisan link.[45]

By 1810, when Isaiah Thomas published a list of American newspapers in his *History of Printing*, 86 percent of American newspapers had a clear party linkage, often in the form of direct financial sponsorship, ad revenues arising from control of government, or promises of future political office.[46] However, unlike the situation at the turn of the nineteenth century when 60 percent of the newspapers supported Federalist candidates, the partisan balance ten years later had become virtually equal. Of the 366 papers across the country, 159 (43 percent) were Federalist, 158 (43 percent) were Republican, and 49 (14 percent) were unaffiliated. According to Lee, most of the "neutral" papers were "agricultural in character" and not devoted to politics.[47]

Newspapers of the day featured several different topics. Primary coverage was devoted to reporting the official proceedings in Washington. Congressional debates were covered in great detail, with policy matters receiving in-depth treatment. Speeches by local representatives in Congress were given significant local play. Newspaper columns were advertised as "always open for communications from politicians of the same political faith."[48] Lest anyone miss the message of the day, long-winded editorials hammered the point home as to what the best course of policy action was for the country. Poems were printed which conveyed political messages and attacks on the opposition.

Similar to his Federalist predecessors, Jefferson continued the trend of supporting an official paper that reported on Washington events. Called the *National Intelligencer* and edited by Samuel Smith, this paper covered the official proceedings of government, from congressional debates to speeches by important dignitaries. According to historical accounts of the paper, "the editors sought to persuade readers to adopt the particular political decisions they reached."[49]

Like Fenno's paper before 1800, Smith's paper largely supported itself through government printing contracts. Starting with Jefferson's administration, the *Intelligencer* was the official printer for the House of Representatives. For the next twenty-six years, it would earn $1 million in government printing contracts and be the dominant house organ. In the days before the creation of the Government Printing Office, these contracts were an indispensable source of revenue for struggling newspapers, many of which would have gone bankrupt except for the political aid.

Smith's devotion to Jefferson was without limit. During the 1800 campaign, the young editor did all he could to help Jefferson win the election. His paper debuted on October 31, 1800, just a few days before the national election. Smith personally wrote a column on the front page proclaiming his support for Jefferson. The rest of the paper was filled with news about official congressional proceedings and a few ads.

After Jefferson won office over Aaron Burr, following a closely contested vote in the House of Representatives, Smith's paper served as Jefferson's administration mouthpiece. While proclaiming the newspaper nonpartisan, the editor dutifully toed the president's line and extolled Jefferson's virtues both on domestic and foreign policy.

Early in Jefferson's term, the *National Intelligencer* got in trouble for printing an anonymous letter complaining about the quality of justice administered by Federalist judges. The local district attorney immediately started libel proceedings against the editor, but the charges were dropped when prosecutors could not identify the actual author of the letter to the editor.[50]

When Jefferson left office, Smith put the paper up for sale. It was purchased by Joseph Gales, Jr., who in conjunction with a business partner, William Seaton, continued Smith's mission of being "the channel through which the federal administration carried its message to the people."[51] The paper was given administration tidbits to publicize before they were generally made known to other outlets. Throughout the Era of Good Feeling, it was reported, "Gales and Seaton shared the confidence of the Republican Party."[52] Not only were they privy to inside political information, the editors earned $650,000 printing the *Annals of Congress* covering the years 1789 to 1824. These publications were the only record of congressional speeches and debates during this period, and were a tremendous source of government patronage for the newspaper. Gales and Seaton started the *Register of Debates* in 1824, which was followed by the *Congressional Globe* in 1834 and the *Congressional Record* in 1873. The latter subsequently emerged as the official legislative publication.[53]

THE JACKSONIAN REVOLUTION

With the decline of the Federalist Party following the Alien and Sedition Acts and the success of Jefferson in cementing his party's hold on the presidency, Jeffersonians dominated the political era until the election of 1824. By this election, the Founding Fathers had disappeared from the scene, and a new generation of leaders was emerging. Four contenders arose to contest the presidency: William Crawford of Georgia, the secretary of the treasury under James Madison; Henry Clay, the Speaker of the House; Senator Andrew Jackson of Tennessee; and John Quincy Adams, Madison's secretary of state.

By the end of the campaign, no one had captured the majority of 132 votes needed in the Electoral College. The individual who had received the largest number of popular votes was Jackson, who also had the greatest number of Electoral

College votes at 99. He was followed by Adams with 84, Crawford with 41, and Clay with 37. Since no one had captured a majority, the election was forced into the House of Representatives, as the Constitution stipulated. The House immediately chose Adams as president on its first ballot, with thirteen of the twenty-four states voting for him. After becoming president, Adams announced that House Speaker Clay would be his secretary of state.

Jackson quickly cried foul. Claiming that the election had been stolen from him through a "corrupt bargain," Jackson's forces began organizing for the 1828 election. One thing that was obvious from the 1824 contest was the need for better organization and better communication with the general public. Jackson instructed his campaign managers to suggest "that the papers of Nashville and the whole State should speak out with moderate but firm disapprobation of this corruption."[54]

In Jackson's mind, the 1824 election had proven that his real power base was state leaders outside of Washington. Members of Congress feared his populist tendencies and promised to go all out to deny him the presidency. If given a choice, they never would support his bid for chief executive. It was clear he needed to mobilize power at the state and local level.

The key, Jackson thought, would be to take the campaign to the people. Previously, campaigns had not been public events, consistent with the suspicions many of the Founders had about the general public. According to the conventional wisdom of the day, elections should involve opinion leaders and political elites more than the public. Jackson decided in 1828 to run the first presidential campaign designed to engage citizens. The effort involved a series of activities aimed at the public, from speeches at mass meetings to circulars, parades, rallies, and demonstrations. It would be Jackson's way to move the country from a republic run by political elites to a democracy that truly involved voters.

Recognizing the need for organization in his campaign, Jackson introduced parties as permanent political organizations. Prior to this time, parties had been short-term organizations designed to contest temporary activities, such as elections. Between elections, parties were not seen as being very important. Jackson's brilliance, along with that of his campaign manager Martin Van Buren, was to turn the temporary organization of parties into an ongoing political force. The value of parties as an organizational weapon was their efficacy in contesting elections and mobilizing the general public. Parties were the perfect organizational device because they provided a way to rise above political elites and appeal to people in states and localities across the country.

But it wasn't just parties that were important. Jackson understood the value of communications channels in public mobilization. Newspapers and circulars were effective ways to communicate views to large groups of people and attract support to the political parties. Much as Hamilton and Jefferson before him had seen the value of newspapers as party organs, Jackson understood the role papers could play in terms of political advocacy. He felt that newspapers should be used as party mouthpieces. Around this time, 50 percent of the contents of metropolitan papers and 70 percent of that of nonmetropolitan papers involved politics, which was a far higher degree of political coverage than is true today when political news pales in comparison to sports, weather, entertainment, and finance.[55]

Jackson was the first presidential candidate to recognize the new role of regional newspaper editors in American politics. He brought into his campaign editors from New England to Louisiana to the Western states, all dedicated to challenging the Washington establishment.[56] One editor even asserted on good authority that Jackson's forces had established a fund of $50,000 in order to build newspaper support for his campaign.[57]

Circulars were used to communicate points of view in starkly partisan and personal terms. For example, in the 1828 elections, a circular for John Quincy Adams attacked Jackson for "ordering executions, massacring Indians, stabbing a Samuel Jackson in the back, murdering one soldier who disobeyed his commands, and hanging three Indians." Not to be outdone, a Jackson circular portrayed Adams as "driving off with a horsewhip a crippled old soldier who dared to speak to him, to ask an alms."[58]

Prior to this time, newspaper reading was an elite activity limited to the wealthy and privileged class. Newspapers were expensive to purchase and readership levels numbered about 2 percent of the population. However, starting with the Jacksonian era, the price of newspapers dropped in order to attract a much larger circulation base. Newspaper subscription rates rose to 3.3 percent in 1850 and 4.7 percent in 1860 (see Appendix Figure A.1). The number of papers increased from 359 to 852 between 1810 and 1828, and the number of copies printed tripled during this period.[59] With the expansion of American territories into the West, the number of papers soon exceeded a thousand.

One of the press innovations during this time that expanded the numbers of papers was the creation of the so-called *penny press*, in which newspapers were sold for a cent (or four dollars annually), down from the previous price of six cents (or ten dollars a year). The ten-dollar subscription was more than most skilled workers earned in a week. Papers previously were not sold on the streets, but required a full year's subscription.[60] Higher circulation rates allowed papers to begin to charge higher advertising rates. As we will see in Chapter 3, the increasing reliance on ad revenues later in the nineteenth century eventually would liberate publishers from party and government patrons and dramatically change the nature of the newspaper business.

The first penny paper, appropriately entitled *Cent*, was issued in 1830 in Philadelphia. Though it failed when its proprietor, Christopher Conwell, died from cholera in 1832, the floodgates were opened. The first permanent penny newspaper was the *New York Sun*. Started on September 3, 1833, this paper boldly announced its mission on the front page, proclaiming: "The object of this paper is to lay before the public, at a price within the means of every one, all the news of the day, and at the same time offer an advantageous medium for advertisements."[61]

The paper was distributed both through subscriptions (three dollars a year) as well as on the street by news carriers, a new innovation for the newspaper business. Yearly advertisements cost thirty dollars and included a complimentary subscription. Within three years of its founding, the *Sun* had a circulation of 27,000 copies daily, which was 20 percent higher than the combined sales of its eleven, six-cent rivals in New York.[62]

In an effort to capture the attention of ordinary people, press coverage in the *Sun* began to include human interest material beyond official proceedings in Washington. The front page of the first issue had stories about ships leaving for and arriving from Europe, the dialogue of an Irish captain who had been in six duels, a feature story about miniature mechanical marvels, and a story about a little boy who whistled so much it almost cost him his life. Page two continued with a story about the suicide of Fred Hall, a twenty-four-year-old gentleman from Boston who was about to sail for Sumatra. The police report outlined offenses ranging from murder and disturbing the peace to wife beating and passing counterfeit money. The last page ended with a series of ads and a poem of fifteen stanzas entitled "A Noon Scene."[63] In general, owing to its desire to reach more of a mass audience, the penny press was less partisan in its coverage of public affairs than its six-cent rivals.

Notwithstanding the gains in readership, journalism was still a low-status occupation. Reporters had little or no formal education and often came from poor backgrounds. For example, in his famous travels around the United States in 1831 and 1832, Alexis de Tocqueville was struck by the low educational levels of American editors. He noted that "The journalists of the United States are usually placed in a very humble position with a scanty education and a vulgar turn of mind."[64]

In reaching out to people and enlisting newspapers and parties in the general mobilization effort, Jackson was very successful. Not only did he win the presidency in 1828 over incumbent John Quincy Adams, voter turnout tripled between 1824 and 1828, from around 400,000 to nearly 1.2 million.

Once in office, Jackson started the process of transforming the American political system into a more democratic union. Endorsing the principle of "rotation in office," he kicked out many Federalist bureaucrats and installed his own supporters, thereby laying the groundwork for the party patronage system. He extended Jeffersonian principles of states' rights, reduced the debt, and improved the nation's transportation infrastructure. Editors of papers that supported him were given major patronage plums in return for favorable coverage.[65]

Jackson continued the tendency of past presidents to have an official paper of the administration. Under John Quincy Adams, the *National Journal* was the favored paper that published the texts of debates in Congress. It was edited by Peter Force, who lived in Washington. Another Adams' outlet was the *Intelligencer*, the Jefferson favorite edited by Joseph Gales, an editor with the pragmatic habit of cultivating close ties with whoever happened to be in office.

Following Jackson's inauguration, a paper called the *United States Telegraph*, edited by Duff Green, became the official paper. Later, after Green demonstrated more loyalty to Jackson's vice president John Calhoun, Jackson set up a new paper, the *Globe*, edited by Francis Blair of Kentucky. Profitable printing contracts were conferred on the paper, and federal employees making more than $1,000 were asked to subscribe to the newspaper.[66] Within a year, the *Globe* had four thousand subscribers and annual government printing contracts worth $50,000, many of which were taken away from the opposition newspaper, the *Telegraph*. This loss caused that paper to cease publication in 1837.[67]

To ensure there was no public confusion over the paper's stance, public speeches by opposition political figures were not printed in the *Globe*. Only news favorable to Jackson was included, and the president often leaked key information first to the *Globe* before notifying other papers. According to Gerald Baldasty, a leading scholar on this period, there was little question during the Jacksonian era that, "the press was at the very center of American political life."[68] This same author concluded that "the highly political Jacksonian era represents what was probably the high tide of the partisan press."[69]

The lead editorial writer at the *Globe* was Amos Kendall. Each night, Jackson and Kendall had a private conference at which the president dictated ideas for future articles. The two would go over possible articles until the items were in the form desired by Jackson.[70] In 1835, near the end of Jackson's presidency, Kendall became postmaster general for the federal government, a patronage reward that was common under Jackson.

One of the ways that Jackson cultivated editors and reporters was to support them for public office. He routinely sent names of editors to the Senate for confirmation to various offices. This practice, though, got him in trouble with his vice president, John Calhoun. Sensing that two editors from a group of four who had been nominated were hostile to the South, Calhoun persuaded the Senate to confirm the two Southern editors and reject the two Northern ones. This set off a lively debate in the Congress as to whether editors even should be eligible for public office in the United States.

By 1830, fifty-six editors and printers had been appointed by Jackson to government positions, a significant portion of his overall appointments.[71] Defending himself against criticism, Jackson justified the practice saying, "Why should this class of citizens be excluded from offices to which others, not more patriotic, nor presenting stronger claims as to qualification, may aspire?"[72]

This obvious conflict of interest with the press became so controversial that in December 1831 the National Republican Party convention drafted a platform plank condemning "corruption of the press" by the Jackson administration. One person noted that Jackson had complained about partisan use of the press in previous administrations, but was engaging in the same practice himself. "Partisan editors were now the most favored class of pretenders to office," that critic said.[73]

THE WEAKNESS OF THE PARTISAN PRESS

Throughout the early decades of the American republic, government officials and party leaders exercised enormous clout over newspaper content because of the financial power of government advertising and printing contracts.[74] Newspapers were kept in line by threats to "stop the Government advertising."[75] Since there were several newspapers in each community, it was easy for public agencies to move their advertising dollars around when there was unfavorable coverage. If threats were not sufficient, direct bribes for party support were "fairly numerous."[76]

One of the ways government officials kept control over papers was through the "list of letters uncalled for," which referred to mail that had not been picked up. Through the postal authorities, ads were placed in newspapers showing who had letters to pick up. It was a major form of patronage for newspapers, and a way for government officials to control press content. For example, when the *New York Evening Post* criticized the postmaster general, the list of uncalled for letters was transferred to a rival New York paper.[77] Later, when the *Post* criticized the secretary of the treasury for an "undignified" tone in a letter to the president, the Treasury Department withheld its advertising from that paper.

Such financial support was absolutely critical to many newspapers. Publishing was a risky business during the first few decades of the country's existence. Subscription levels were not very high and ad revenues were modest by contemporary standards. According to one study, two-thirds of newspapers established between 1815 and 1836 failed in their first few years.[78]

It was an era that afforded great power to public officials who controlled government contracts. Writing about a leading journalist in the 1830s named James Gordon Bennett, biographer Isaac Clark Pray noted: "Newspapers then were an expensive luxury, owned and supported by politicians or sectarists who found it to their interest to invest money even in losing speculation, and who deemed their hired editors to be the convenient tools of caprice and pleasure, while the public was a simple multitude to be cajoled and deceived on every subject."[79]

In the relationship that existed between reporters and politicians, public officials saw themselves as much more important than reporters. When dignitaries gave speeches, they saw the major audience as those actually attending the event, not people at a distance who could be cultivated through press coverage. Once, for example, Henry Clay was about to make a speech in Kentucky, when he was told that an Associated Press reporter named Richard Smith from Cincinnati was present to cover the event. Clay refused to give the speech because the journalist had not requested special permission to report Clay's thoughts to the general public. Smith was forced to leave the event and was only able to learn the speech's highlights from a friend after its delivery.[80] Such behavior stands in stark contrast to the media ascendancy that would emerge in the twentieth century.

Despite the blatant partisanship of press coverage during this era, few were concerned about its impact on voters. The large number of news outlets, the heterogeneity of the coverage, the low public esteem toward the press, and the obvious partisan leanings of publishers limited the power of the press to be influential. Reporters were not renowned for fairness and accuracy in their coverage so readers took news reports in the spirit in which they were written. Either people read only the papers that conformed to their previously established views (which limited the ability of reporters to alter opinions) or they read multiple papers in order to get a balanced view of what was happening (which restricted the power of any single newspaper).

From a normative point of view, outside observers did not worry much about the deleterious consequences of press power or tabloid-style reporting. In their view, the low cost and diversity of media outlets provided protection against vitriolic coverage from any single newspaper. As a European traveler to the United States, distinguished geologist Charles Lyell of Great Britain noted "the cheapness of the innumerable daily and weekly papers enables every villager to read what is said on more than one side of each question, and this has a tendency to make the multitude think for themselves, and become well informed on public affairs."[81] Early American newspapers might be outlandish and slanted in their coverage, but such reports did not have much influence on public opinion or voting behavior.

NOTES

1. Clinton Rossiter, ed., *The Federalist Papers* by Alexander Hamilton, James Madison, and John Jay (New York: Penguin Books, 1961).
2. Thomas Slaughter, *The Whiskey Rebellion* (New York: Oxford University Press, 1986).
3. Rossiter, *The Federalist Papers*.
4. *The Gazette of the United States*, April 15, 1789. Also see George Henry Payne, *History of Journalism in the United States* (New York: D. Appleton, 1920), 154; and Robert W. Jones, *Journalism in the United States* (New York: E. P. Dutton, 1947), 175.
5. James Melvin Lee, *History of American Journalism* (Boston: Houghton Mifflin, 1917), 163.
6. William Dill, "Growth of Newspapers in the United States," University of Kansas Bulletin, 1928, 11.
7. Lee, *History of American Journalism*, 122; Culver H. Smith, *The Press, Politics, and Patronage: The American Government's Use of Newspapers 1789–1875* (Athens: University of Georgia Press, 1977), 13; and Timothy E. Cook, *Governing with the News: The News Media as a Political Institution* (Chicago: University of Chicago Press, 1998), 25–26.
8. Henry Randall, *The Life of Thomas Jefferson* (Philadelphia: J. B. Lippincott, 1888), 1: 560. Also see Payne, *History of Journalism in the United States*, 157; and Claude Bowers, *Jefferson and Hamilton* (Boston: Houghton Mifflin, 1925), 17.
9. Lee, *History of American Journalism*, 122.
10. *National Gazette*, December 19, 1791. Also see Frank Luther Mott, *American Journalism: A History of Newspapers in the United States through 260 Years: 1690 to 1950*, rev. ed. (New York: MacMillan, 1950), 124. For full-length treatments of Freneau, see Jacob Axelrad, *Philip Freneau: Champion of Democracy* (Austin: University of Texas Press, 1967); Samuel E. Forman, *The Political Activities of Phil Freneau*, vol. 20 (Baltimore: Johns Hopkins University Press, 1902); and Lewis Leary, *That Rascal Freneau* (New Brunswick, N.J.: Rutgers University Press, 1941).
11. Thomas Jefferson, *Writings of Thomas Jefferson* (Monticello edition), Andrew Lipscomb, ed., (Washington, D.C.: Thomas Jefferson Memorial Association, 1903–4), 5: 181–182. Also see Payne, *History of Journalism in the United States*, 157; Bowers, *Jefferson and Hamilton*, and James Pollard, *The Presidents and the Press* (New York: Macmillan, 1947).
12. Lee, *History of American Journalism*, 123.
13. *Gazette of the United States*, June 6, 1792.
14. John Tebbel and Sarah Miles Watts, *The Press and the Presidency: From George Washington to Ronald Reagan* (New York: Oxford University Press, 1985), 13.
15. Mott, *American Journalism: A History of Newspapers in the United States*, 125.
16. Jones, *Journalism in the United States*, 170. Also see William David Sloan, "Scurrility and the Party Press, 1789–1816," *American Journalism*, 1988.
17. Payne, *History of Journalism in the United States*, 159.
18. Frederic Hudson, *Journalism in the United States, From 1690 to 1872* (New York: Harper & Brothers, 1873), 181. Also see Worthington Ford, *Jefferson and the Newspaper, 1785–1830* (New York: Columbia University Press, 1936).
19. Paul Leicester Ford, ed., *Writings of Thomas Jefferson* (New York: Putnam, 1892–1899), 1: 231. Also see Hudson, *Journalism in the United States, From 1690 to 1872*, 186; and Payne, *History of Journalism in the United States*, 166.
20. Lee, *History of American Journalism*, 128.
21. Mott, *American Journalism*, 124.
22. Bowers, *Jefferson and Hamilton*.

23. John C. Miller, *Crisis in Freedom: The Alien and Sedition Acts* (Boston: Little, Brown, 1951), 8.

24. Mott, *American Journalism*, 147.

25. Miller, *Crisis in Freedom: The Alien and Sedition Acts*, 8.

26. See Alien Act of July 6, 1798, 1 Stat. 596, *Annals of Congress*, 5th Cong., 2d sess.

27. See Sedition Act of July 14, 1798, 1 Stat. 596, *Annals of Congress*, 5th Cong., 2d sess.

28. Hudson, *Journalism in the United States, From 1690 to 1872*, 210ff. For material on Bache, see Benjamin Franklin Bache, *Truth Will Out! The Foul Charges of the Tories against the Editor of the Aurora Repelled by Positive Proof and Plain Truth and his Base Culumniators Put to Shame* (Philadelphia: Aurora, 1798); and James Tugg, *Benjamin Franklin Bache and the "Philadelphia Aurora"* (Philadelphia: University of Pennsylvania Press, 1991).

29. Mott, *American Journalism*, 128. Also see Payne, *History of Journalism in the United States*, 174.

30. Bowers, *Jefferson and Hamilton*, chapter XVII.

31. Payne, *History of Journalism in the United States*, 180–181.

32. Jones, *Journalism in the United States*, 171.

33. Mott, *American Journalism*, 149. Also see Payne, *History of Journalism in the United States*, 179–180.

34. Mott, *American Journalism*, 152.

35. Payne, *History of Journalism in the United States*, 195.

36. Mott, *American Journalism*, 169.

37. The joke is cited in Saul K. Padover, ed., *Thomas Jefferson on Democracy* (New York: Mentor, 1960), 97. Also see Tebbel and Watts, *The Press and the Presidency*, and Mott, *Jefferson and the Press*.

38. Michael Emery and Edwin Emery, *The Press and America: An Interpretive History of the Mass Media*, 8th ed. (Boston: Allyn & Bacon, 1996), 77.

39. Ibid., 75.

40. Mary Voboril, "Sex Scandal and Presidency: History Repeats Itself," *Miami Herald*, January 10, 1999.

41. Ibid. For an account of the Jefferson-Hemings relationship, see Annette Gordon-Reed, *Thomas Jefferson and Sally Hemings: An American Controversy* (Charlottesville: University of Virginia Press, 1997).

42. Payne, *History of Journalism in the United States*, 196. Also see Jerry Knudson, "Political Journalism in the Age of Jefferson," *Journalism History* (Spring 1974).

43. Court case *People v. Croswell*, 3 Johns. (N.Y.) 377 (1804). Also see Payne, *History of Journalism in the United States*, 197; and Tebbel and Watts, *The Press and the Presidency*, 36.

44. Emery and Emery, *The Press and America*, 79.

45. One exception to this newspaper pattern of partisan connections was *The Manufacturers' and Farmers' Journal and Providence and Pawtucket Advertiser*, which started printing on January 3, 1820, and later in 1829 became a daily paper known as the *Providence Journal*. Although not affiliated with any political party at its inception, it did have a clear point of view — its advocacy of high tariffs to protect American industry. It was the official outlet of the Rhode Island Society for the Encouragement of Domestic Industries. See Lee, *History of American Journalism*, 146.

46. Isaiah Thomas, *History of Printing in America* (Worcester, Mass.: Isaiah Thomas, Jr., 1810).

47. Lee, *History of American Journalism*, 117.

48. Ibid., 145.

49. William Ames, *A History of the National Intelligencer* (Chapel Hill: University of North Carolina Press, 1972), viii. Also see Emery and Emery, *The Press and America*, 85.

50. Ames, *A History of the National Intelligencer*, 28–29.

51. Ibid., 88.

52. Ibid., 111.

53. Emery and Emery, *The Press and America*, 85.

54. Tebbel and Watts, *The Press and the Presidency*, 77. Also see Robert Remini, *Andrew Jackson and the Course of American Democracy, 1833–1845* (New York: Harper & Row, 1984).

55. Gerald J. Baldasty, *The Commercialization of News in the Nineteenth Century* (Madison: University of Wisconsin Press, 1992), 23.

56. Tebbel and Watts, *The Press and the Presidency*, 78.

57. Ibid., 77.

58. Kathleen Hall Jamieson, *Packaging the Presidency*, 3rd ed. (New York: Oxford University Press, 1996), 6–7.

59. Tebbel and Watts, *The Press and the Presidency*, 75.

60. Payne, *History of Journalism in the United States*, 240–241.

61. Jones, *Journalism in the United States*, 228.

62. Payne, *History of Journalism in the United States*, 244–246.

63. Jones, *Journalism in the United States*, 232.
64. The first quote comes from Payne, *History of Journalism in the United States*, 241; the second is from Mott, *American Journalism: A History of Newspapers in the United States*, 168.
65. Tebbel and Watts, *The Press and the Presidency*, 81.
66. Ibid., 82.
67. Mott, *American Journalism: A History of Newspapers in the United States*, 179.
68. Gerald J. Baldasty, "The Press and Politics in the Age of Jackson," *Journalism Monographs* LXXXIX (August 1984): 23.
69. Baldasty, *The Commercialization of News in the Nineteenth Century*, 5.
70. William Stickney, ed., *Autobiography of Amos Kendall* (Boston, 1872). Also see Hudson, *Journalism in the United States, From 1690 to 1872*, 238–239; and Payne, *History of Journalism in the United States*, 235.
71. Baldasty, *The Commercialization of News in the Nineteenth Century*, 20.
72. Tebbel and Watts, *The Press and the Presidency*, 85–86.
73. Payne, *History of Journalism in the United States*, 259.
74. Timothy E. Cook, *Governing with the News: The News Media as a Political Institution* (Chicago: University of Chicago Press, 1998), chapters 2 and 3.
75. Lee, *History of American Journalism*, 223.
76. Ibid.
77. Ibid.
78. Baldasty, *The Commercialization of News in the Nineteenth Century*, 19.
79. Quoted in Dill, "Growth of Newspapers in the United States," 6–7.
80. Lee, *History of American Journalism*, 215–216.
81. Charles Lyell, *A Second Visit to the United States of North America* (London: John Murray, 1849), I: 60–61.

CHAPTER 3

The Commercial Media

On May 24, 1844, Professor Samuel Morse of New York University made history by sending the first telegraph message — "What hath God wrought?" — from Washington, D.C. to Baltimore. Sitting in the old Supreme Court chambers, Morse conveyed this information using commercial wires that ran the forty miles between the two cities on posts twenty feet high and one hundred yards apart.[1] Working out of classrooms in the old University Building on Washington Square in New York, the professor had been experimenting with an electric device that instantaneously sent messages over wires.

The same day that he made inventor's history, Morse delivered the first telegraph message published in a paper, the *Baltimore Patriot*. Referring to a congressional action that afternoon, the inventor typed: "One o'clock — There has just been made a motion in the House to go into committee of the whole on the Oregon question. Rejected — ayes, 79; nays, 86."[2]

Taking advantage of this new innovation along with advances in high-speed printing technology and cheaper newsprint, reporters began relying on the telegraph to transmit news items that previously had been transmitted by mail. For example, in 1844, the telegraph was used to send a message from the Democratic national convention at Baltimore to Silas Wright in Washington, D.C., informing him he had been nominated for vice president. Wright replied via telegraph that he was not interested. However, the convention adjourned for a day awaiting confirmation because it did not believe the telegraph message.[3]

In 1845, James Gordon Bennett, one of the most prominent editors of his day, forecast the tremendous impact the telegraph would have on American political life. In a *New York Herald* editorial, he predicted the telegraph "will be more influential than ever. The public mind will be stimulated to greater activity by the rapid circulation of news."[4] Referring to the new invention, he said it "communicates with the rapidity of lightning from one point to another. The whole nation is thus impressed with the same idea at the same moment."[5] Famed New York newspaper editor Horace Greeley echoed this sentiment when he predicted to Morse, "You are going to turn the newspaper office upside down with your invention." By 1848, more than five thousand miles of wire had been installed across America.[6]

The telegraph would become one of many technologies and industrial developments that would commercialize news gathering and help build the American nation. Between technological advances, improvements in printing, the industrialization of America, and the extension of penny newspapers to the middle class, a commercial revolution took place in the media. Subscription levels rose from 2.6 percent of Americans in 1840 to 19.8 percent in 1900 (see Appendix Figure A.1), which attracted businesses interested in reaching this new mass consumer audience. By 1900, advertising revenues comprised the majority of income for American newspapers.

The rise in subscribers and advertisers liberated publishers and editors from the power of party leaders and government officials, and gave them financial incentives to listen to their customers. National interest in events such as the Civil War and Reconstruction heightened the need to write stories that would be transmitted to readers all across America, which dictated more homogeneous coverage. Papers still maintained a partisan edge in their coverage, but the reason was due to commercial needs more than direct party sponsorship of the newspaper.[7]

By the start of the twentieth century, the commercial pressures of this era pushed reporters into a cutthroat tabloid competition for readers. Freed of political patrons, a series of strong publishers and editors attempted to outdo one another in attracting subscribers and advertisers. These competitive pressures eventually ushered in a tabloid style of coverage known as *yellow journalism*, which built circulation levels but did not enhance the public credibility of reporters. Journalists came to be seen as sensationalistic and fixated on crime and vice, which limited the ability of the press to influence public opinion on political matters. It was hard for readers to take press accounts very seriously when the media industry was so strongly motivated by commercial pressures.

THE IMPACT OF TECHNOLOGY ON NEWS TRANSMISSION

The immediate impact of the telegraph was in transforming the scale and speed of news gathering. Because it made possible the joint sharing and communication of information, Morse's invention encouraged news pooling by different media organizations across the country. Ironically, it was a foreign policy event, the 1846 Mexican War, that demonstrated this new trend. Starting with that war, the *New York Herald* and the *Tribune* ran virtually identical news stories gleaned from the telegraph. This led to a formal decision in 1848 and 1849 to create a New York–based organization called the Associated Press, which became the nation's first wire service. By 1880, Simeon North reports that 228 of 438 morning papers (52 percent) and 127 of 533 evening papers (24 percent) were relying on telegraphic news.[8]

With the development of the telegraph in 1844, the rise of wire services, and national expansion of the transportation system through roads and the railroad, technology had the effect of broadening the perspective taken by many journalists and contributing to a new national self-image. Several national dailies became prominent during this period, which aided the cross-country dissemination of information and the process of nation building.

Transportation long had been a limiting factor in news dissemination. In the colonial period, for example, the major mode of transportation was via rivers and highways. Before 1812, there were 37,000 miles of road in the United States, but transportation was never easy. When the Cumberland Pike cost too much to build, the contractor was urged to go faster and leave tree stumps in the right of way, but to be careful that "the stump did not protrude more than a foot above the ground."[9]

By 1850, there were 9,021 miles of railroad in twenty-six states. Coupled with the growing use of the steam engine and the construction of new canals such as the Erie Canal and the Chesapeake and Ohio canal, railroads speeded up movement across the country. The expanding rail network had a dramatic effect on information communication. According to one writer, "Rail transportation made possible the early delivery of newspapers, even to the interior of the country."[10]

At the end of the nineteenth century, the country was tied together through an interconnected industrial, communications, and transportation network. Railroad lines grew to 93,000 miles in 1880 and 193,000 in 1900. Free rural delivery of the mail by the federal government came into operation in 1897.[11] A new Alexander Graham Bell invention known as the telephone was making Americans even less remote from one another. From its origin in 1876, the telephone reached 1 percent of American households by 1900 and many more soon thereafter (see Appendix Figure A.2). Each of these technological and political developments proved to be a tremendous boon to the process of nation building that characterized the United States in the last half of the nineteenth century and early part of the twentieth century. They also made it possible for a series of new editors to gain power, build circulation levels, and transform the media industry.

THE RISE OF INDEPENDENT EDITORS

Around the time these improvements in transportation and communication were being made, the news business was changing perceptibly. Prior to this, many editors saw their job as serving a political party and acting as the mouthpiece for particular leaders. Through their control of government contracts and advertising, public officials ensured that editors were financially dependent on political patronage. It would not be until the creation of the Government Printing Office in 1860 that direct printing subsidies to the partisan press would officially end.[12]

The rise of more commercially oriented newspaper publishers and editors, however, altered the relationship between the press and political officials. These media executives were more independent of partisan political forces, more nationally oriented, and more interested in gaining commercial advantage within the industry. Few individuals illustrated these trends more than James Gordon Bennett of the *New York Herald*, Horace Greeley of the *New York Tribune*, and Henry Jarvis Raymond of the *New York Daily Times* (later shortened to the *New York Times*).

Bennett was a British national who had immigrated to America in 1819 at the age of twenty with less than $25 to his name. After working as a news reporter on a Charleston ship, Bennett became the Washington correspondent for the *Enquirer*. Covering the 1828 presidential election, he ardently supported Andrew Jackson and his populist reforms to extend democracy in America.[13] Bennett then went on to establish his own paper, the *New York Globe* in 1832, and the *Pennsylvanian*, both of which died for lack of popular support.

On May 6, 1835, though, Bennett hit pay dirt when he started the *Herald*. The newspaper was virtually a one-man operation, with Bennett doing his own reporting, writing, posting books, and making out bills. It soon would transform the news business.[14] While other papers emphasized politics, the *Herald* was the first successful paper to shift focus to the hot new areas of scandal, crime, and vice. From 20,000 in 1836, the paper rose to a circulation of 77,000 by 1860, making it the world's largest daily paper.[15] Bennett realized that many working class Americans did not care for politics, but wanted to know the lurid details of trials and scandals. He demonstrated a lucrative commercial market for nonpolitical coverage. As such, he was the most prominent forerunner of the tabloids of today.

Part of what made Bennett successful was a flamboyant personal style that attracted both loyalists and critics. One of the things Bennett did was to attack the editors of other papers, many of whom were stock speculators whose news coverage was designed to raise the value of their financial holdings.[16] Nothing pleased Bennett more than attacking bank barons, Wall Street financiers, and his fellow editors. Referring to such editors, Bennett wrote that these men were "truly unfit by nature and want of capacity to come to a right conclusion upon any subject. . . . They pervert every public event from its proper hue and coloring, to raise one stock and depress another. There is no truth in them."[17]

Even former friends were not spared from attack. Once, Bennett wrote a story critical of his former boss, James Watson Webb of the *New York Courier and Enquirer*. Webb was not amused. The competing editor waited for him on the street after the story appeared and beat Bennett with a stick. Demonstrating a remarkable flair for turning lemons into lemonade, Bennett wrote a story for his readers describing the attack. He detailed how Webb had, "by going up behind me, cut a slash in my head about one and one-half inches in length and through the integuments of the skull. . . . He did not succeed, however, in rifling me of my ideas."[18] Following this encounter, the *Herald*'s circulation rose another nine thousand copies.

When Bennett decided to get married, he announced the occasion with bold headlines in his newspaper: "To the Readers of the Herald — Declaration of Love — Caught at Last — Going to be Married — New Movement in Civilization." In the accompanying story, Bennett bared his soul: "I am going to be married in a few days." He described his betrothed as "one of the most remarkable, accomplished, and beautiful young women of the age." Bennett closed by pledging that "the holy estate of wedlock will only increase my desire to be still more useful."[19]

Between Bennett and his son, who followed in his stead, New York had one of the most remarkable set of newsmen who ever lived. Whereas the father revolutionized news gathering by sensationalizing the news, often by gleaning provocative tidbits from police reports, scandals, and gossip, his son created the notion of "exclusive" stories that ran only in his paper. Through such franchise reports, their paper boosted readership and created a loyal following.

One example of Bennett's sensationalism is evident in his coverage of the murder of a prostitute by a man well-known around town. Bennett's paper created so much interest in the case that a courtroom hearing could not begin. Writing about the spectacle, the *Herald* noted, "The Judges and the Officers left the hall. Robinson [the man in question] was carried out of court, and the Public Authorities were trying to clear the hall of the mob."[20]

Across town, Horace Greeley was working a different kind of magic with the *New York Tribune*. Born in New Hampshire in 1811, Greeley started as a newspaper apprentice in Vermont and served on a variety of papers in Pennsylvania. By the time he arrived in New York, he had only $10 in his pocket.[21] After editing a political journal called the *Log Cabin*, which helped William Henry Harrison and John Tyler win the 1840 election, Greeley bought the *Tribune* on April 10, 1841.[22]

The paper was devoted to combating what Greeley saw as a dangerous new trend in American life — the type of newspaper sensationalism practiced by his rival, Bennett. Writing in his very first issue, Greeley described his paper as "a new journal of politics, literature, and general intelligence," which would "advance the interests of the people, and promote their moral, social, and political well being."[23] Lest anyone mistake the target of his effort, Greeley openly disavowed the type of coverage practiced by Bennett. The *Tribune* editor joyfully pronounced that "The immoral and degrading police reports, advertisements, and other matter which have been allowed to disgrace the columns of our leading penny papers will be carefully excluded."[24]

Greeley sought to distance himself from what he thought was the greatest sin of earlier editors — clear and direct political partisanship. Writing in his autobiography, *Recollections of a Busy Life*, Greeley described his goal in starting the *Tribune*: "My leading idea was the establishment of a journal removed alike from servile partisanship on the one hand and from gagged, mincing neutrality on the other. . . . I believed there was a happy medium between these extremes."[25]

Greeley was one of the leading intellectuals of his period. As a sign of his willingness to consider new ideas, he opened his paper to socialists of his era. The wretched winter of 1837–1838 exposed the editor to great personal suffering and had a dramatic impact on his thinking. Writing of that winter, Greeley said, "I saw two families, including six or eight children, burrowing in one cellar under a stable — a prey to famine on the one hand, and to vermin and cutaneous maladies on the other."[26]

This focus did not endear Greeley to Bennett. The *Herald* editor described the *Tribune* as a "socialistic phalanx" in which all of "the editors, printers, publishers, reporters, all the way from the nigger to the lesser devils, are all interested."[27] Seeing himself as a progressive voice, Greeley devoted significant attention to new social movements. In 1848, his paper gave considerable coverage to the first convention of the Women's Rights movement, which was organized in Seneca Falls, New York. Later, the *Tribune* would take on the greatest challenge of Greeley's era — slavery — and in the process, become the most important mass circulation antislavery journal of the era.

Greeley pioneered a new journalistic innovation, the official interview with great men of the day, with a printed verbatim transcript. One example was with Brigham Young, the head of the Church of Jesus Christ of Latter Day Saints. That interview ran in the *Tribune* on August 20, 1859. In the interview, Greeley immediately went to the central issue of the day: "H. G.: What is the position of your Church with respect to slavery? B.Y.: We consider it of Divine institution, and not to be abolished until the curse pronounced on Ham shall have been removed from his descendants."[28]

Henry Jarvis Raymond, the editor of the *Times*, had apprenticed under Greeley. Born in upstate New York, Raymond was not nearly as eccentric as Greeley or Bennett. He had graduated from the University of Vermont in 1840 and gone to work first on Greeley's weekly paper, the *New Yorker*, and then his daily paper, the *Tribune*. While there, he saw the remarkable financial success of the *Tribune*. At a point when it was estimated that the paper was clearing over $75,000 a year, young Raymond was making just $10 a week as an assistant editor and chief reporter. Feeling terribly underpaid, he saw a new paper as a business opportunity rather than a political vehicle.

On September 18, 1851, the *New York Daily Times* became a reality.[29] In its prospectus, the new editor described the paper's mission ". . . to print the local news of the day, insert correspondence from European countries, give full reports of Congressional and legislative proceedings, review books, and contain criticism of music, drama, painting, and any form of art which might merit attention."[30]

Raymond wanted a paper that "avoided Bennett's crudities and Greeley's advocacy of intellectual fads, such as socialism." Raymond felt there was a market for a moderate paper that would appeal to people leaving the Whig and Democratic parties, but who were not yet ready for the strong antislavery views of Greeley's *Tribune*.[31]

That type of political centrism was consistent with Raymond's own personality. Writing the paper's very first editorial, he explained: "There are very few things in this world which it is worthwhile to get angry about. And they are just the things that anger will not improve."[32]

Soon, the *New York Daily Times* was well along the way of becoming a leading national paper. Its circulation and advertising revenue soared under Raymond's guidance. Although active in politics as a member of Congress, he did not use the newspaper as a personal platform the way other editors did. Rather, his goal was to "publish the news," not merely to "print the political views of its editor."[33]

In financial terms, Raymond's investment was a huge success. Aided by the paper's selection by the New York State Banking Department in Albany as the paper in which weekly financial statements were published, the newspaper's stock value climbed exponentially. Starting at around $1,000, the value of the paper rose quickly to $11,000 and then skyrocketed to $1 million.[34] The newspaper was on its way to becoming the paper of record for news coverage.

THE CHALLENGE OF SLAVERY

Few issues illustrate the growing national orientation of press coverage and the powerful role of independent editors in the nineteenth century better than slavery, the Civil War, and Reconstruction. Slavery had been the object of newspaper discussion for a number of years in specialized abolitionist journals, but until the 1850s did not receive a lot of coverage by the mainstream press. For example, William Lloyd Garrison started a small paper known as the *Liberator* on January 1, 1831, but it never reached a subscription of more than three thousand. A loud and impassioned critic of slavery, Garrison defended his fanatical style by arguing that "slavery will not be overthrown without excitement."[35] Many of his journalistic contemporaries, though, did not like the ideological fervor with which Garrison argued his cause.

Nor was antislavery sentiment limited to white editors. There were nearly half a million free blacks in the United States by 1850. Between 1827 and 1865, forty black newspapers addressed slavery and other issues. The first black-published newspaper known as *Freedom's Journal* appeared on March 16, 1827, edited by John Russwurm, the first black man to graduate from an American university (Bowdoin in 1826) and Reverend Samuel Cornish of New York. Among other items, this paper railed against "the inhumanity of slavery."[36]

Frederick Douglass also gained fame during this era for starting an antislavery paper called *The North Star* in Rochester, New York, that was printed from 1847 to 1860. Featuring articles by black writers, this paper established a reputation for literary excellence. The newspaper was financed with money raised through abolitionist lectures around the world and sales from his best-selling autobiography, *Life and Times of Frederick Douglass*.[37]

But aside from specialty publications with small circulations, slavery did not take center stage in the newspaper business until the Kansas-Nebraska Act was enacted. The Missouri Compromise of 1820 had temporarily quieted discussion of this contentious national issue with its agreement to admit Maine as a free state and Missouri as a slave state, and to make all of the Western territory land north of Missouri free soil. While the decades following this decision would witness periodic skirmishes between contending sectional forces, the debate did not engage the national psyche or attract much attention from white-run newspapers.

All this changed in 1854, though, when Democratic Senator Stephen Douglas introduced and Congress passed the Kansas-Nebraska Act. This legislation split the Nebraska Territory in two (Kansas and Nebraska) and permitted inhabitants of those states to decide for themselves, through the doctrine of "popular sovereignty," whether slavery should be permitted in their lands. More so than any other single action, this law reactivated an issue that had been lurking in the background of American politics and thrust it once again on center stage.

Many newspapers around the country denounced the legislation.[38] Since it repealed the Missouri Compromise and created a mechanism for every new state to introduce slavery by a vote of the people, antislavery forces saw it as a clear vehicle for the expansion of slavery to Western territory. Leading the charge against the bill was the prominent editor Greeley. In his mind, the Kansas-Nebraska Act represented everything that was wrong with American politics on the slavery issue. To Greeley, the compromise was immoral and also wrong politically; by inflaming public passions and making possible an increase in slavery, the legislation did not move the country in the right direction.

Following the lead of its editor Greeley, the *New York Tribune* argued on January 6, 1854, that the "Missouri Compromise declared that neither slavery nor involuntary servitude shall ever exist in the Northwest Territories. Stephen Douglas's Nebraska bill sets a precedent to override that declaration and to allow slavery above the 36 degrees 30 minutes line." It called the act "a breach of solemn compact between the North and the South" and forcefully denounced "every attempt to remove the salutary restriction upon the introduction of Slavery into the North-West."[39]

The *Washington D.C. Union* replied nine days later, calling the *Tribune* "an organ of abolitionism," driven by its "usual fanatical bitterness." "The course of the *Tribune*," stated the *Union*, "more than ever confirms us in the importance which we attach to the Nebraska report and bill." The paper continued to say that if the act did not pass with its "popular sovereignty" clause, it "would deny the territory's people the right of self-government."

The *Tribune* answered this argument two days later, citing the act as an attempt by the South to bring slavery to the North. Passage of the legislation risked future hostilities, the paper said. If enacted, "the whole strength of the North will be brought into the field against this infamous project . . . Sober minded men, who have leaned to the side of the South . . . will turn and resist this movement as a gross outrage and aggression on the part of the South."

It is little surprise that papers from the North and South had very different impressions of what was happening. The *Charleston (S.C.) Mercury*, for example, accused the North of a "determination to set at naught [any] provisions, in every respect, at all favorable to the South." The paper complained that the Missouri Compromise was simply "a hollow truce by which the South was put to sleep for further robbery."

The *Hartford (Conn.) Courant* meanwhile accused Douglas of starkly political motives. Sponsorship of the bill, it said, was simply a "bold bid of Douglas for the next Presidency." Continuing, the *Courant* noted: "We shudder at the danger to which the country has already been exposed from its existence. But we see a perfect fire brand flung into the political arena for the aggrandizement of one man. The measures of the Compromise have been settled and the nation is going on peaceably and prosperously under their operation. Why renew then the agitation?"

In an effort to rally support, Stephen Douglas himself delivered a speech a few days later, arguing that any attempt to decide the slavery issue without input of settlers would shake the foundations of American democracy. Several Southern newspapers, including the *Mercury*, came out with editorials supporting Douglas's position, while Northern journals redoubled their opposition. Articles in the *New Orleans Bulletin*, "the most influential Whig journal in the South," the *Louisville Journal*, "the leading Whig paper in the West," and the *National Intelligencer* voiced dissatisfaction with the act.

Rather than quieting the national debate, the Kansas-Nebraska Act moved the slavery controversy back into the forefront of national dialogue. All sides of the slavery battle realized this legislation upped the ante on the divisive issue. Slavery become a topic that would engage every newspaper editor in the country. Eventually, it would engulf the nation in a civil war.

At a convention in Saratoga, New York, on August 16, 1854, Greeley proposed that a new Republican party be formed. Made up of Whigs, Free Soilers, and anti-Nebraska Act Democrats, the new party would strive to outlaw the nefarious practice of slavery.[40] With newspaper subscriptions of more than 200,000, much of it in the West, Greeley was in a strong position to push his favorite cause.[41]

On the other side was Greeley's old nemesis Bennett. The *New York Herald* endorsed the Kansas-Nebraska Act and Bennett described his paper as "the only Northern journal that has unfailingly vindicated the constitutional rights of the South."[42] This, of course, did not endear Bennett to the abolitionists, but it made him very popular with Southern political leaders.

In the Midwest, the *Chicago Tribune* edited by Joseph Medill took a firm stance against slavery. Started in 1847, this paper became a major backer of a young politician named Abraham Lincoln.[43] As the leading paper in the Northwest Territory, the *Tribune* dominated political discussions in Illinois, Iowa, Wisconsin, and Minnesota.[44] Medill had been impressed with Lincoln ever since he heard the orator deliver an impassioned speech demanding the abolition of slavery. The young man understood the moral imperative raised by the issue, Medill thought. After hearing the speech, Medill ceaselessly promoted Lincoln to all who would listen.[45]

As the press argued about slavery, public officials debated the issue as well. In one prominent congressional exchange, Senator George Badger of North Carolina stated, "If some Southern gentleman wishes to take the old woman who nursed him in childhood and whom he called 'Mammy' into one of these new territories for the betterment of the fortunes of his whole family — why, in the name of God, should anybody prevent it?" Senator Benjamin Wade of Ohio replied, "We have not in the least objection to the Senator's migrating to Kansas and taking his old 'Mammy' along with him. We only insist that he shall not be empowered to sell her after taking her there."[46]

When the key congressional vote on the Kansas-Nebraska Act occurred, it followed geographical and party lines. In the Senate, the bill was approved 37-14. The House's vote was much closer with 113 for and 100 against the bill. Overall, 61 percent of the yes votes in the House were Southern, while 91 percent of the no votes were from Northerners.[47]

Following passage, Southern papers such as the *Charleston Mercury* and the *Milledgeville (Ga.) Federal Union* published glowing tributes to Douglas. Northern papers, however, touted the act's victory as an incident that "enabled the slave States to fling off the mask and show what their intentions and determinations are. These plans are two-fold, relating to both the internal and external condition of the country. So far as the acquisition of foreign territory is concerned, their next great step is the seizure of Cuba. This acquisition will add to the number and wealth of the slave States and furnish an additional market for slave-raising Virginia."[48]

Opponents called on the North to rally against further expansion of slavery. The next year was characterized by "Bleeding Kansas," an armed conflict between antislavery and proslavery mobs that lasted from the summers of 1854 to 1859. This conflict would become the opening shot in a longer term drama — before long, a full-fledged Civil War was under way between the North and the South. Coverage of that event would bring major changes to the newspaper industry, including extensive reliance on the telegraph and new printing technologies, nationally syndicated wire reports, and editors who saw their mission as encouraging politicians to take particular positions on major issues.

THE CIVIL WAR

Perhaps no election saw newspaper editors play as controversial a role as occurred in the 1860 presidential election. Taking place at the most tumultuous time in the nation's history, with sectional interests in direct collision and slavery roiling the ability of politicians to respond, this election brought a new leader, Lincoln, to the presidency and pushed the United States into a bloody and tumultuous Civil War.

From the standpoint of Lincoln, the crucial editor in this election was Greeley. As a long-time champion of the Whig Party, Greeley had been expected to support William Seward for president over Lincoln. Indeed, Greeley, Seward, and an editor named Thurlow Weed of the *Albany (N.Y.) Evening Journal* had a long political association with one another. Seward was the clear front-runner for the nomination and an individual who was respected in the East.

However, when the convention met in Chicago, with support from Greeley, Lincoln surprised the political experts and wrestled the nomination away from Seward. The Illinois politician then went on to win the presidential general election in a four-way contest with Stephen Douglas of the Northern wing of the Democratic Party, James Breckenridge of the Southern wing of the Democratic Party, and John Bell of the Union Party.[49] Lincoln received 1.8 million votes, followed by Douglas with 1.3 million, Breckenridge with 850,082 votes, and Bell with 646,124.

Over the course of several years, Seward had committed a number of blunders that upset Greeley. For one, he had opposed Greeley's idea of a new Republican party. Following the counsel of Thurlow Weed, Seward did not believe a new party was needed. It was better, he thought, to work within existing parties. In Greeley's eyes, this argument was seriously flawed; such a position demonstrated that Seward lacked the personal strength and leadership necessary to fight the slavery battle.

Even worse, Weed and Seward had denied party support of Greeley when he wanted to run for statewide office in New York. Greeley had called upon Weed as the titular head of the old Whig Party in New York for support as a gubernatorial candidate. Weed replied that he "did think the time and circumstances favorable to his election, if nominated, but that my friends had lost control of the state convention" to the Know-Nothings, an anti-immigrant and anti-Catholic group that had become powerful in the 1840s and 1850s. When pressed by Greeley for help in running for lieutenant governor, Weed offered little assistance. To make matters even worse from Greeley's standpoint, the convention ended up nominating his newspaper rival Henry Raymond for the post.[50]

Shortly thereafter, Greeley started casting about for a new candidate. In 1858, Lincoln and Stephen Douglas held their famous series of U.S. Senate campaign debates in Illinois, proclaiming their respective positions. In these encounters,

Douglas supported the "popular sovereignty" position on slavery that paralleled the one presented in the Kansas-Nebraska Act. Slavery should be decided by what the voting inhabitants of each state wanted, he argued; or to put it differently, slavery was a political issue to be decided by referendum, not a moral controversy.

Lincoln took a strong stance against this view and uttered his famous prediction that "a house divided against itself cannot stand."[51] Although he lost the Senate race, Lincoln became a presidential candidate in 1860. Campaigning in New York, Greeley met Lincoln and came away dazzled by the Illinois Republican. In his paper, Greeley wrote referring to Lincoln as "one of nature's orators."[52] At the convention, Greeley shocked Seward by casting his lot with Lincoln and helping to deliver a number of Western states to his candidacy. On November 7, 1860, Lincoln won the general election and became president. In an effort at reconciliation with a party opponent, Lincoln brought Seward into his cabinet as secretary of state.

It was not a propitious time to be elected president given national tensions about slavery and states' rights. Lincoln's election on an antislavery platform stunned the South — it was not an outcome Southerners had thought possible. On November 10, right after Lincoln triumphed in the popular vote (but before the Electoral College had met), the South Carolina legislature voted to call a December 20th convention to consider the possibility of secession. This followed an editorial campaign in the *Charleston Mercury* blasting the federal government for trampling on states' rights. In fact, Robert Rhett, the editor of that newspaper, had been one of the earliest people to advocate secession as a Southern policy. At the South Carolina convention, delegates voted unanimously to secede.

Northern newspapers and many politicians took the view that the U.S. Constitution in no way interfered with the right of *existing* states to maintain slavery. Speaking on March 4, 1861, in his inaugural address, Lincoln followed this line of reasoning. He noted that with his election, "apprehension seems to exist among the people of the Southern states that by the accession of a Republican administration their property and their peace and personal security are to be endangered." Seeking to assuage those anxieties, Lincoln said, "I have no purpose, directly or indirectly, to interfere with the institution of slavery in the States where it exists. I believe I have no lawful right to do so, and I have no inclination to do so."[53] His only goal was to stop the spread of slavery into new territories. That was where he wanted to draw the line. But the effort at reconciliation was to no avail. Within months, the country would be at war with itself.

The Civil War had an enormous impact on the mass media. Every newspaper in the country used war correspondents at the front to cover the conflict. Headlines boldly proclaimed the latest change in fortunes for each side, based on telegraph reports from the front lines. Because telegraph lines occasionally were cut, reporters developed the "news lead," in which the most important development was summarized at the front of the story. It was a way of making sure readers got the most important point of the day in the event communications lines failed.

In addition, local papers reported the list of war casualities. As *the* story of national interest across the country, such articles were devoured with great intensity by those on the home front. Newspapers were the way many family members learned that loves ones had been killed or injured in the war. Distribution of the local paper, therefore, was a big event. Depending on the news reported, it became a time either of great joy or deep tragedy in many communities across America. The news greatly affected the personal fortunes of a good many people.

With improvements in communications and the need to get news out fast to all parts of the country, news became syndicated. Everything from commentary to sports, fiction, letters from the lovelorn, comic strips, puzzles, and market information was distributed by national networks of news disseminators. Previously, each locale and each region had its own editors and writers who were famous in that area; now, syndication created national figures. Regardless of whether one was in New York, Chicago, or St. Louis, it was possible to read the same story by the same writer of what was happening in the world.[54]

Syndication was a way for small papers to pool resources, distribute entertaining and informative material to readers, and gain economies of scale in the printing process. With the loss of many men who specialized in newspaper typesetting to the war effort, syndicates enabled papers in small communities to stay in business. For example, in 1861, Ansell Kellogg published a weekly newspaper in Wisconsin. When his compositor enlisted in the Army, Kellogg was unable to print a full-sized newspaper. He therefore made arrangements for the editors of the *Wisconsin State Journal* to send him "half-sheet supplements filled with war news, to be folded in with his weekly paper." These were a big time-saver for many small newspapers. Within two years, this cooperation had grown to thirty papers in the state, and by 1866, it included sixty-five newspapers across Wisconsin, Michigan, Minnesota, Illinois, Ohio, Kansas, and Tennessee.[55]

From the standpoint of the federal government, the most difficult war issue was reconciling freedom of the press with national security interests. On August 5, 1861, General McClellan met with war correspondents about what was "suitable" to include in news coverage. His fears that Northern papers would report the strength and movements of his troops soon proved very real. While Southern papers almost never reported the movement of Confederate troops, their Northern counterparts did, thereby negating the possible advantage of surprise attacks. McClellan publicly complained that information was being printed that would "furnish aid or comfort to the enemy."[56]

Northern leaders suppressed newspapers which violated national security rules on troop movements. At other times, they placed papers under the supervision of friendly war correspondents. Direct censorship was adopted when generals saw fit. A censor was placed in the Washington telegraph office to inspect all news dispatches.[57] False news reports were filed in order to deceive the Confederacy.

Because of the public's appetite for news and soaring increases in circulation levels, papers spent heavily on war coverage. For example, the *Herald* spent half a million dollars on its war correspondents. Reporters were sent to the far reaches of the conflict in order to file timely reports. According to one historian of the era, "every army of the North had its *Herald* headquarters equipped with tents, a wagon bearing the name of the paper, and several attendants."[58]

Increasingly concerned about Lincoln, in 1862 Greeley wrote an editorial entitled, "The Prayer of Twenty Millions." In it, he said that he was "sorely disappointed and deeply pained" at the president's unwillingness to do more for slaves. Within a week, Lincoln replied publicly. He announced in regard to Greeley's fears that "my paramount objective in this struggle is to save the Union and is not either to save or destroy Slavery. If I could save the Union without freeing any slave, I would do it."[59]

As the conflict turned in favor of the Union, Northern papers filled their columns with sympathetic stories about the war. Human interest articles were printed about heroic efforts on the battlefield. Official communiqués ran verbatim. The details of every battle were told and retold.

In the end, the national government triumphed and restored the Union. The last proclamation of Jefferson Davis, president of the Confederacy, was published April 5, 1865, when Richmond fell to the North. Reprinted widely throughout the North, Davis conceded that "it would be unwise to conceal the moral and material injury to our cause resulting from the occupation of our capital by the enemy."[60]

Soon, the war would end and a president would be assassinated. Given the changes taking place nationally in the media, it was fitting that on April 14, 1865, the first journalist to report Lincoln's shooting was Lawrence Gobright of the Associated Press. Signaling the important role the AP was playing nationally, his message tersely reported "the President was shot in a theater tonight, and perhaps mortally wounded."[61]

RECONSTRUCTION

The end of the Civil War left the country sharply divided and undergoing change on a massive scale. The North was left prosperous while the South was devastated politically and economically. Most of the war had been fought on Southern soil, and it would take generations for the damage to heal. According to some estimates, the South lost $5.8 billion in slave property, staple crops, war devastation, and Confederate debt.[62]

Fueled by the technological imperative of the conflict, a vast Industrial Revolution expanded and swept across the country. National rail networks were created, and new communications channels built. A new national union emerged, vastly different from the Jeffersonian republic that had existed in 1800.

The immediate task following cessation of hostilities was what to do about the South. Eleven states had seceded and gone to war with the national government. Following the Union victory, Northern leaders wanted the South punished for secession. But debates raged over the severity and form of this punishment.

President Andrew Johnson urged leniency toward the South. Not only did he oppose restoring liberty to slaves, he tacitly supported so-called *black codes* adopted by Southern states that restricted "the movement, property rights, and individual liberties of former slaves."[63] Over the objections of Congress, he unilaterally tried to readmit Confederate states along with their established leadership back into the Union without major preconditions guaranteeing the constitutional rights of slaves.

Many Northern politicians protested Johnson's policy. From their standpoint, provisions needed to be enacted that would safeguard civil rights and provide for public instruction of all people, regardless of color. Reflecting that view, in 1865, the Thirteenth Amendment was added to the U.S. Constitution abolishing slavery and involuntary servitude. Three years later, over Johnson's objections, the Fourteenth Amendment guaranteed equal protection under the law. Its language made it illegal for any state to "deprive any person of life, liberty, or property, without due process of law; nor deny to any person within its jurisdiction the equal protection of the laws." In 1870, the Fifteenth Amendment guaranteed the right of male suffrage to all regardless of "race, color, or previous condition of servitude."

Ultimately, the political divisions unleashed by Johnson's policy would lead to his near removal from office in 1868 on impeachment charges. The close vote demonstrated how narrowly divided the country was at this point in time. The president's weakness in the aftermath of impeachment laid the groundwork for control by Congress. Relying on their new political power, "radical Republicans" installed Northern "carpetbaggers" throughout the South. With conservatives in retreat and blacks voting in substantial numbers, these officials ran the region as a conquered territory.[64] Republicans hoped to use new black voters to cement party control over the entire region.

Southern states were forced to apply for readmission to the Union on terms favorable to the North. Civil rights legislation extended democratic rights to former male slaves, including the right to run for office. By 1868, six Southern states had complied with these demands and were readmitted to the Union. The remaining states were in full compliance by 1870.[65]

In the short run, Reconstruction appeared to be successful at restoring civil rights and extending democratic privileges. Many blacks ran for office and a record number of black officeholders were elected throughout the South. African Americans won places in every state legislature in the South and actually gained a majority of seats (eighty-seven to sixty-nine seats) in the lower chamber of the South Carolina legislature.[66] In Mississippi, there was a black lieutenant governor, secretary of state, superintendent of education, and Speaker of the House. Louisiana had a black lieutenant governor, secretary of state, treasurer, and superintendent of public education. Florida had a black secretary of state and superintendent for public instruction. South Carolina had a black lieutenant governor, secretary of state, treasurer, Speaker of the House, and associate justice of the state supreme court. In all, fourteen blacks were elected to the U.S. House of Representatives and two served in the U.S. Senate.[67] It appeared that Reconstruction was a smashing success and that not only slavery had been ended, but that African Americans were using the vote to gain greater political equality.

But these advances soon proved illusory. Below the surface, the war had destabilized the economies both of the North and South. In the panic of 1873, the economy came crashing down, deflating prices and incomes throughout the country. There was insufficient currency to fund the expanding nation. Farmers, workers, and poor people who owed money were destroyed overnight. In short order, cash to finance the moral crusade against the South dried up. According to the U.S. Bureau of the Census, the rate of business failure in the United States during the 1870s reached a higher point than it later would obtain in the 1930s at the time of the Great Depression.[68]

In this situation, not only did the South recede as the object of national discontent, political protest arose against Eastern money interests in the form of the Populist Party and the Greenback Party. Each party saw an Eastern establishment that was pursuing its own interests at the expense of the South and West.

With commercial interests at stake, newspapers began to take a more moderate stance on Reconstruction. The *New York Times*, for example, supported a middle-of-the-road perspective on black voting rights. It supported black suffrage, but argued that "suffrage should be extended by constitutional amendment to all men *who could read* (emphasis added), regardless of color."[69] This argument provided a crucial loophole through which white Southern conservatives later could deny the right to vote to blacks who failed to pass a literacy test.

By 1874, a number of Southern states had drifted back to the Democratic Party, the bastion of white conservative voters. Republicans had lost control of three-quarters of Southern state legislatures.[70] The balance of political power also shifted in the U.S. Congress. While Republicans retained the Senate, the split between Republicans and Democrats in the House was nearly even. The South became solidly Democratic. Eighty of the 107 Southern House members had served in the Confederate army.[71]

This trend was evident in the 1876 presidential election. The close contest between Democrat Samuel Tilden and Republican Rutherford Hayes signaled the end of Reconstruction. Neither man gained a majority due to contested vote totals in Louisiana, Florida, and South Carolina. Congress was forced to convene an extraordinary Electoral Commission to resolve the dispute. Although Tilden had won a larger proportion of the popular vote, the Electoral Commission decided in favor of Hayes. In his March 5, 1877, inaugural address, Hayes argued that local autonomy ought to be restored to the Southern states.[72] His decision to appease the South sounded the death knell for Reconstruction.

Following these election results, most of the gains won by blacks were lost. Home rule was restored to the South in 1877. Northern interlocutors left the region to return home. The number of federal troops left in the South dropped to three thousand.[73] The minuscule federal effort to enforce voting rights ended altogether. In the end, Hayes's Southern strategy effectively ended the federal government's liberal experiment on race.

These decisions only confirmed what already had been acknowledged as the de facto reality. During this period, the record of federal election enforcement was quite weak. According to one leading Reconstruction historian, "only 34 percent of the cases tried under the enforcement laws in the South between 1870 and 1877 resulted in conviction, and a mere 6 percent of them in the border states."[74]

With this pitiable record, it was not long before whites' latent hostility towards blacks reemerged. White vigilante groups arose in every Southern state. Using violence and lynchings, these groups intimidated blacks into silence and non-participation. From 1882 to 1930, it has been estimated that 2,805 victims were lynched in ten Southern states. While some of these victims were white (around 300), the vast majority (nearly 2,500) were black. Ninety-four percent of blacks who were murdered died at the hands of white lynch mobs.[75]

It was not just lynchings that rolled back the progress of Reconstruction. According to scholars, "beatings, whippings, verbal humiliations, threats, harangues, and other countless indignities" were perpetuated on the black population. As the hand of federal regulation receded, Southern states passed laws restricting black employment. For example, in South Carolina, blacks had to pay a tax of up to $100 if they worked in an occupation other than farmer or servant. Mississippi passed a law making it illegal for any free black man to keep firearms or knives. Whites were forbidden to sell or give guns, knives, or ammunition to blacks.[76]

Black voter turnout plummeted. Most black elected officials who had gained office after the Civil War lost their positions of power. By the late nineteenth century, many blacks had been reduced to economic and political impoverishment.

Throughout this period of violence and repression, some white newspapers reported the lynchings almost as a spectator sport. But the black press spoke out forcefully against the violence. The most influential black newspaper, the *Chicago Defender*, attacked white oppression and the lynching of blacks. The *Richmond Planet*, a paper founded in 1883 as a "voice to Virginia's black residents," denounced lynchings and called for action against them.[77]

A courageous black female journalist named Ida Bell Wells Barnett organized an antilynching crusade in the 1890s. After three of her friends were lynched in 1892, she wrote an editorial denouncing the crimes. Shortly thereafter, her newspaper, the *Memphis Free Speech*, was destroyed by local whites. Bravely, she began a crusade to investigate lynchings, traveling throughout the United States and England to lecture about the topic. Her 1895 book, *A Red Record*, provided one of the most detailed accounts of black lynchings.[78]

Some Northern papers also condemned the lynchings and printed graphic descriptions of the torture and murder of black victims. On June 11, 1900, for example, the *New York Times* published a story about an innocent man in Mississippi being lynched. On December 7, 1899, a *New York World* story recounted the burning alive of Richard Coleman for the alleged murder of a white farmer's wife. However, the *New York World* later covered the speech of Professor Albert Bushnell Hart of Harvard University before the American Historical Association saying that "if the people of certain States are determined to burn colored men at the stake, those States would better legalize the practice."[79]

Other papers had coverage which framed the lynchings in ways that subtly justified the murders. This, of course, undermined the moral force of their condemnations. When the first black cadet at West Point was taken from his bed, bound, beaten, and cut, the *New York Truth Seeker* reported on April 17, 1880, that "the other cadets claim that he did it himself." On April 28, 1899, The *Kissimmee Valley (Fla.) Gazette* described the torture and burning of a black man named Sam Holt at a stake, surrounded by a mob of two thousand. In graphic detail, the paper noted that at one point the man, who had been accused of murdering a white man and raping a white woman, "admitted his guilt."[80]

A Southern reporter visiting Maine provided a chilling first-hand account to the *Bangor Commercial* on September 5, 1899, describing the practical problem of covering a lynching. A number of editors sent reporters to cover lynchings, thereby treating them as just another news event. In somber language, the reporter described what it was like being the journalist on hand to write the lynching article. According to the story, he said, "The news that there is to be a lynching spreads very rapidly in the south, especially in the small cities and towns. To the reporter it is a very disagreeable business to attend these lynchings, for he is usually not overcome by frenzy like the mob made up from the immediate neighborhood." The writer noted that he had covered twelve lynchings, some where the person seemed guilty and others where they did not.[81] Although the confession captured the moral ambivalence felt by the reporter, this story also demonstrated the subtle and not-so-subtle ways in which newspapers contributed to this tragic aspect of Reconstruction.

THE EMERGENCE OF PULITZER AND HEARST

With the end of Reconstruction, the growing concentration of political and financial power in the hands of the few unleashed an even stronger national role for journalists. By a decade following the end of the Civil War, the great editors of the 1840s, 1850s, and 1860s were dead: Greeley of the *Tribune*, Bennett of the *Herald*, and Raymond of the *Times*, among others. In their place emerged new media leaders such as Joseph Pulitzer and William Randolph Hearst, who would guide journalism into an era of cutthroat commercial competition and tabloid coverage.

Befitting an era characterized by sometimes crass commercialism, there was big money to be made in the newspaper business. As shown in Appendix Figure A.1, newspaper circulations rose from 4.7 percent in 1860 to 19.8 percent in 1900. A new middle-class market was opening up for the media. There also was a tremendous expansion in the country's wealth during this period. Between 1865 and 1880, national wealth doubled. Over the last two decades of the nineteenth century, it doubled again. According to Baldasty, "by 1900, advertising revenues accounted for as much as two-thirds of all newspaper revenues."[82]

Born in Hungary in 1847, Joseph Pulitzer became one of the most important men in commercializing the media. Starting with a newspaper in St. Louis, the *Dispatch* (later merged with the *Post*), Pulitzer continued the trends that had begun with his predecessors. Buying the *Dispatch* for $2,500 in 1878 at a bankruptcy sale, Pulitzer launched a revamped paper that within four years was clearing $45,000 a year.[83]

His editorial policy was one of independence. Stating his credo for the paper, he wrote: "The *Post* and *Dispatch* will serve no party but the people; be no organ of Republicanism, but the organ of truth; will follow no causes but its conclusions; will not support the 'Administration,' but criticize it; will oppose all frauds and shams wherever and whatever they are; will advocate principles and ideas rather than prejudices and partisanship."[84]

But political independence did not mean blandness. Pulitzer loved to crusade against crooked politicians and the wealthy on behalf of the middle class. One of his discoveries was that there were a lot of working-class people who were not reading papers. Between his attacks on corruption and juicy gossip about leading families in the area, Pulitzer found a way to draw those people to newspapers.[85] Critics referred to his approach as tabloid sensationalism, but the new style of coverage also had one noble feature. It succeeded in broadening the circulation basis of newspapers around the turn of the century.

With his St. Louis paper bringing him vast wealth, Pulitzer decided to enter the New York market. In 1883, he purchased the *New York World*, and within four years this newspaper broke every record in the media business for circulation and profitability. By the end of his first year, circulation was more than 60,000; by 1887, it had risen to 250,000.[86]

The *World* combined the two elements that had been so successful in St. Louis: sensationalism and progressive crusades. In 1893, it earned the nickname "yellow journalism" when it issued its first colored supplement featuring a cartoon called "The Yellow Kid" meant to publicize the newspaper.[87] Soon, critics were referring to Pulitzer's brand of sensational coverage, big headlines, and colorful graphics as yellow journalism.

Pulitzer was not apologetic about the tone of his papers. Explaining his approach, he said, "The American people want something terse, forcible, picturesque, striking, something that will arrest their attention, enlist their sympathy, arouse their indignation, stimulate their imagination, convince their reason, [and] awaken their conscience."[88]

About the same time Pulitzer was transforming his New York newspaper, William Randolph Hearst was becoming interested in journalism. Hearst's father was a multimillionaire who had made his fortune on gold mines in South Dakota and New Mexico. The Hearst family also owned dozens of ranches from Alaska to Mexico.[89] Enrolled at Harvard University in 1882, Hearst did not take his studies very seriously. His class attendance was irregular and more than one course was dropped in mid-semester when he found the instructor too boring. Occasionally, Hearst left school in the middle of the term to travel with his parents. He also got into trouble frequently by playing practical jokes and partying. One famous incident occurred when he sent "a jackass into one of his professor's rooms with an attached card that read: 'Now there are two of you.'" Another time, he threw custard pies at people in the Howard Athenaeum. Within three years, he suffered the ultimate embarrassment — expulsion from the university.[90]

Since his major interest was journalism, he asked his father for help. Hearst senior had taken over the *San Francisco Examiner* as payment for a bad debt. Hearst told his father, "I want the *San Francisco Examiner*." His father exploded: "Great God! Haven't I spent money enough on that paper already? I took it for a bad debt and it's a sure loser."[91] Instead, his wealthy father offered him a ranch in Mexico. The son declined. Realizing he needed greater experience in the newspaper business to impress his father, Hearst went to New York and started working as a reporter for Pulitzer's *World* paper. The experience left an indelible mark on Hearst. In three years, Pulitzer had raised the circulation from 15,000 to 250,000 and was making an enormous profit.

Hearst got his big break in 1887 when his father was chosen by the California Legislature as U.S. senator. Leaving California, Senator Hearst gave his son the gift he had lobbied so long for — the editorship of the *San Francisco Examiner*. Young

Hearst was twenty-three years old at the time. The new editor hired a professional staff and started the task of turning the paper around. His father had lost $250,000 on the paper, but his strategy had been seriously flawed. Hearst's father blatantly used the paper as a personal organ to advance his own political career. The son, though, had different plans. Borrowing ideas from Pulitzer's *New York World,* Hearst made the *San Francisco Examiner* more colorful. He put sports news, such as baseball, on the front page, which boosted public interest. He used bold headlines and made shameless claims that the paper was the "LARGEST, BRIGHTEST AND BEST NEWSPAPER ON THE PACIFIC COAST."

Hearst used periodic promotions to grab readers' attention. One day, he provided a newspaper coupon good for a free, seven-hour cruise on the Bay. The next day, four thousand people showed up for the free boat ride. Since the ship only had room for one thousand people, three thousand were turned away disappointed.[92]

Realizing that big news did not happen every day, Hearst sought to create major stories that would attract readers and advertisers. He ran a series of community crusades against a new city charter, high water rates, and the Southern Pacific Railroad, which was infamous for exorbitant rates and poor service. When the charter was defeated and water rates reduced by 16 percent, he claimed personal credit on behalf of regular folks.

By the end of Hearst's first year, the *Examiner's* circulation was up from 24,000 to 40,000 and advertising revenues doubled, though the paper still had an annual deficit of $300,000. But soon, the red ink stopped and the paper was making a yearly profit of $500,000.[93]

By 1895, Hearst had enough money to buy the *Morning Journal* in New York. Again imitating the *World* format, Hearst introduced splashy tabloid features on crime and scandal. He also raided the *World* for its best writers and editors, which made Pulitzer furious. From a meager 77,000, circulation rose to over 100,000 readers.

As a result of the head-to-head competition between Hearst and his former boss Pulitzer, the two papers became the leading practitioners of yellow journalism. Each paper competed hard with the other for the most sensationalistic news and gossipy coverage. Hearst was not disturbed by the tone of his paper. Defending its coverage, he proclaimed, "the public is even more fond of entertainment than it is of information."[94]

THE SPANISH-AMERICAN WAR OF 1898

Nothing crystallized yellow journalism and the new emerging national and international role of the United States more than the Spanish-American War in 1898. With jingoistic headlines from leading papers, journalists fanned the flames of war and encouraged America to go to war to assert national prerogatives. Indeed, it was during this conflict that banner headlines set in all capital letters first were used in American papers, a technique that Hearst pioneered.

The first reporters on the scene were dubious about prospects for a war. Sent by Hearst to draw the war, famed illustrator Frederic Remington wired home the often-quoted remark: "There will be no war. I wish to return." Wanting to beat Pulitzer's paper, Hearst instructed the illustrator to stay where he was. "You furnish the pictures and I will furnish the war," he boldly proclaimed.[95] Hearst quickly sought to turn this claim into a reality.

Each day, after Hearst and other leading journalists had decided a war was inevitable, American papers were filled with accounts of Spanish atrocities in Cuba. Stories about Spanish troops burning huts and imprisoning peasants were given prominent coverage. Accounts of women being strip-searched by Spanish forces were featured. For example, the story of one young Cuban female named Clemencia Arango who was subjected to this treatment was played up in the *New York Journal*, with the story accompanied by a risqué Remington illustration designed to draw attention to the article.[96] Similar accounts of barbarism by Cuban rebels were not covered. As in many wartime situations, atrocities by fighters sympathetic to the American side avoided detection. Not only was coverage tilted in favor of natives against the Spanish, reporters bribed Cuban rebel leaders with medicine and rum to gain access to exclusive stories.

On February 15, 1898, when the USS *Maine* was blown up in Havana harbor, the press frenzy was in full force. Although evidence presented at a congressional investigation later suggested the sinking was an accident arising from problems in the ship's own ammunition magazine, American reporters blamed the disaster on the Spanish. Bold headlines claimed the Spanish enemy had sunk an American battleship.

Editorially, Hearst pushed elected officials to go to war with Spain. President William McKinley was reluctant, fearing the war would damage the American economy. But with public fears whipped into a frenzy by the press coverage, the United States declared war on Spain and sent troops to Cuba, Puerto Rico, and the Philippines. War historian Allan Keller concluded based on archival research that "had these publishing titans not decided to slug it out toe to toe, the efforts of the downtrodden Cubans to throw off the yoke of Spanish oppression might never have burgeoned into a war between Spain and the United States."[97]

After American forces triumphed over the Spanish fleet and captured Manila on August 14, the *New York Journal* used its largest headline ever to proclaim, "MANILA OURS!"[98] By the time the war ended five months after it started, the United States had become an acknowledged world leader. In recognition of its global power, America now had new territories scattered around the world.

THE MEDIA AND NATION BUILDING AT THE TURN OF THE TWENTIETH CENTURY

Between the Civil War and the Spanish-American War, the American media had become a different entity. Befitting broader social, economic, and political changes, technological developments, and the key role of powerful editors and publishers, the media evolved into a commercially oriented industry featuring tabloid stories. Newspapers relied on subscribers and advertisers for revenue more than government printing contracts or political patrons.

The Civil War broadened the outlook of many reporters and produced stories that were national in scope. The existence of local troops being sent to places around the country created heightened interest in national news coverage. Every hamlet avidly sought news from far away about the course of the war and fate of hometown boys.

The Spanish-American War further widened the horizons of reporters, news coverage, and readers. Reporters used the considerable public interest in the war to win new readers and to advance the doctrine of American intervention around the world. Slowly but steadily, the country moved from an isolated hamlet guarded by two oceans to a unified nation with global aspirations.

In the process, new media barons emerged. Both Joseph Pulitzer and William Randolph Hearst grew quite wealthy. With no income tax on the books and generous inheritance rules on passing possessions across generations, rich men dominated the newspaper industry. By the early twentieth century, Hearst had a national media empire of unimagined magnitude: twenty-five daily papers, seventeen Sunday papers, four syndicates, one newsreel, and thirteen magazines, among others, at an estimated value of $220 million.[99]

Overall, chains captured 31 percent of newspaper circulation by 1923, and 42 percent by 1935.[100] The parochialism of the partisan press had been broken once and for all. Editors had advanced beyond their roots in a newspaper system dominated by government officials and party leaders. By the early twentieth century, the American media were more commercially oriented and national in their style of coverage. Except for isolated situations, journalists still did not have the power they would gain later in the twentieth century. However, as they threw off their role as partisan mouthpieces, journalists were moving toward a more powerful role in the political system. But the tabloid coverage and sensationalistic headlines did little to boost their public credibility. Readers were as skeptical of papers dominated by commercial interests as those run for the benefit of blatantly political objectives. It would take another media transformation before reporters would gain the full trust and respect of the American public.

NOTES

1. William Dill, "Growth of Newspapers in the United States," University of Kansas Bulletin, 1928, 53; Robert W. Jones, *Journalism in the United States* (New York: E. P. Dutton, 1947), 252, and Michael Emery and Edwin Emery, *The Press and America: An Interpretive History of the Mass Media*, 8th ed. (Boston: Allyn & Bacon, 1996), 114.

2. Emery and Emery, *The Press and America*, 114.

3. Jones, *Journalism in the United States*, 252.

4. Dill, "Growth of Newspapers in the United States," 53.

5. Ibid., 53; Alvin Harlow, *Old Wires and New Waves: The History of the Telegraph, Telephone, and Wirless* (New York: Appleton-Century-Crofts, 1936); and Menachem Blondheim, *News Over the Wires: The Telegraph and the Flow of Public Information in America, 1844–1897* (Cambridge, Mass.: Harvard University Press, 1994).

6. The Greeley quote comes from James Melvin Lee, *History of American Journalism* (Boston: Houghton Mifflin, 1917), 274. The wire mileage is noted in Charles Lyell, *A Second Visit to the United States of North America* (London: John Murray, 1849), 1: 243.

7. Gerald J. Baldasty, *The Commercialization of News in the Nineteenth Century* (Madison: University of Wisconsin Press, 1992); and Michael McGerr, *The Decline of Popular Politics: The American North, 1865–1928* (New York: Oxford University Press, 1986).

8. Simeon North, *History and Present Condition of the Newspaper and Periodical Press of the United States* (Washington, D.C.: U. S. Government Printing Office, 1884). A discussion of the history of the telegraph is found in Emery and Emery, *The Press and America*, 115; and Donald L. Shaw, "News Bias and the Telegraph: A Study of Historical Change," *Journalism Quarterly* 44 (1967): 3–12, 31.

9. Dill, "Growth of Newspapers in the United States," 51.

10. Ibid., 53.

11. Emery and Emery, *The Press and America*, 158.

12. Richard M. Perloff, *Political Communication* (Mahwah, N.J.: Lawrence Erlbaum, 1998), 22.

13. George Henry Payne, *History of Journalism in the United States* (New York: D. Appleton, 1920), 257–258. Also see Don Seitz, *The James Gordon Bennetts* (Indianapolis: Bobbs-Merrill, 1935); Anonymous, *The Life and Writings of James Gordon Bennett* (New York, 1844); and Isaac Clark Pray, *Memoirs of James Gordon Bennett* (New York: Stringer & Townsend, 1855).

14. Payne, *History of Journalism in the United States*, 260–261.

15. Emery and Emery, *The Press and America*, 103.

16. Payne, *History of Journalism in the United States*, 261.

17. Ibid.

18. Ibid., 262.

19. Ibid., 265–266.

20. Emery and Emery, *The Press and America*, 103.

21. Payne, *History of Journalism in the United States*, 272–273. Also see William Cornell, *Life of Horace Greeley* (Boston: Lee & Spehard, 1872); and W. A. Linn, *Horace Greeley* (New York: D. Appleton, 1903).

22. Payne, *History of Journalism in the United States*, 277.

23. Jones, *Journalism in the United States*, 267.

24. Ibid.

25. Ibid., 277–278. Also see Horace Greeley, *Recollections of a Busy Life* (New York: J. B. Ford, 1868).

26. Payne, *History of Journalism in the United States*, 279.

27. Jones, *Journalism in the United States*, 269.

28. Ibid., 273–274.

29. Lee, *History of American Journalism*, 270.

30. Ibid., 271. Also see Augustus Maverick, *Henry J. Raymond and the New York Times* (Hartford, Conn.: Hale & Co., 1870).

31. Dorothy Dodd, "Henry J. Raymond and the *New York Times* During Reconstruction," (Ph.D. dissertation, University of Chicago, 1936), 5.

32. *New York Times* editorial, September 18, 1851. Quoted in Maverick, *Henry J. Raymond and the New York Times*, 98.

33. Lee, *History of American Journalism*, 271. A contemporary history of the *New York Times* is presented in Susan Tifft and Alex Jones, *The Trust: The Private and Powerful Family behind the New York Times* (Boston: Little, Brown, 1999).

34. Lee, *History of American Journalism*, 272–273.

35. Emery and Emery, *The Press and America*, 125.

36. Ibid., 126–127.

37. Frederick Douglass, *Life and Times of Frederick Douglass* (Hartford: Park Publishers, 1882).

38. Lee, *History of American Journalism*, 279–280.

39. Quotes from newspapers in the following paragraphs come from the Furman University Department of History's Secession Era Editorials Project (Summer 1998), <http://www.furman.edu/~ benson/docs/editorial/>.

40. Payne, *History of Journalism in the United States*, 286–287.

41. Emery and Emery, *The Press and America*, 130.

42. Ibid., 131.

43. Ibid., 131–133.

44. Jones, *Journalism in the United States*, 264.

45. Perloff, *Political Communication*, 22.

46. Congressional debate quotes from Furman University Department of History's Secession Era Project (Summer 1998), <http://www.furman.edu/~benson/docs/>. Also see James Malin, *The Nebraska Question, 1852–1854* (Lawrence: University of Kansas Press, 1953).

47. Congressional vote from Furman University Department of History's Secession Era Project (Summer 1998), <http://www.furman.edu/~benson/docs/>.

48. Quotes from Furman University Department of History's Secession Era Editorials Project (Summer 1998), <http://www.furman.edu/~benson/docs/editorial/>.

49. Payne, *History of Journalism in the United States*, 282–292; Jones, *Journalism in the United States*, 281; and Lee, *History of American Journalism*, 207–210.

50. Payne, *History of Journalism in the United States*, 288–289.

51. Ibid., 292.

52. Ibid., 293.

53. Jones, *Journalism in the United States*, 309.

54. Ibid., 350–354.

55. Ibid., 352–354.

56. Lee, *History of American Journalism*, 288–289.

57. Ibid., 291.

58. Ibid., 293.

59. Ibid., 295.

60. Jones, *Journalism in the United States*, 321.

61. John Tebbel and Sarah Miles Watts, *The Press and the Presidency: From George Washington to Ronald Reagan* (New York: Oxford University Press, 1985), 201.

62. Dodd, "Henry J. Raymond and the *New York Times* during Reconstruction," 25. Also see *New York Times* editorial, June 27, 1865.

63. Joshua Zeitz, "Impeachment: Johnson Deserved It; Clinton Doesn't," *George St. Journal*, February, 1999, 12; and Eric Foner, *Reconstruction: America's Unfinished Revolution, 1863–1877* (New York: Harper & Row, 1988).

64. Emery and Emery, *The Press and America*, 147.

65. Kenneth Stampp, *The Era of Reconstruction, 1865–1877* (New York: Knopf, 1966), 145.

66. Ibid., 167; and Foner, *Reconstruction: America's Unfinished Revolution*.

67. Stampp, *The Era of Reconstruction*, 167.

68. Stewart Tolnay and E. M. Beck, *A Festival of Violence: An Analysis of Southern Lynchings, 1882–1930* (Urbana: University of Illinois Press, 1995), 12. Also see Mark Wahlgren Summers, *The Press Gang: Newspapers & Politics, 1865–1878* (Chapel Hill: University of North Carolina Press, 1994).

69. Dodd, "Henry J. Raymond and the *New York Times* During Reconstruction," 28. Also see *New York Times* editorials, January 4, 1864; December 29, 1864; June 19, 1865; and June 21, 1865.

70. William Gillette, *Retreat from Reconstruction 1869–1879* (Baton Rouge: Louisiana State University Press, 1979), 166.

71. Ibid., 251; and Foner, *Reconstruction: America's Unfinished Revolution*.

72. William Gillette, *Retreat from Reconstruction 1869–1879*, 337.

73. Ibid., 35.

74. Ibid., 42.

75. Tolnay and Beck, *A Festival of Violence: An Analysis of Southern Lynchings*, ix, 271–272.

76. Ibid., ix, 4.

77. Ellen Nakashima, "Afro-American Paper Dies," *Washington Post*, February 8, 1996, B3.

78. *Encyclopedia Britannica Online*, "Ida Bell Wells Barnett," (Summer 1998), <http://www.eb.com>.

79. Ralph Ginzburg, *100 Years of Lynchings* (Baltimore: Black Classic Press, 1962), 24, 31, 36.

80. Ibid., 9, 11.

81. Ibid., 21.

82. Emery and Emery, *The Press and America*, 156. The Baldasty quote comes from Gerald Baldasty, "The Nineteenth-Century Origins of Modern American Journalism," in *Three Hundred Years of the American Newspaper*, John Hench, ed. (Worcester, Mass.: American Antiquarian Society, 1991), 409.

83. Emery and Emery, *The Press and America*, 173. Also see Don Seitz, *Joseph Pulitzer* (Garden City, N.Y.: Garden City Publishing, 1924); and Alleyne Ireland, *Joseph Pulitzer* (New York: M. Kennerly, 1914).

84. Emery and Emery, *The Press and America*, 174.

85. Payne, *History of Journalism in the United States*, 360–369.

86. Emery and Emery, *The Press and America*, 177.

87. Jones, *Journalism in the United States*, 413.

88. Ibid., 424.

89. John Winkler, *William Randolph Hearst* (New York: Simon & Schuster, 1928), 32–35.

90. Ben Proctor, *William Randolph Hearst: The Early Years 1863–1910* (Oxford: Oxford University Press, 1998), 32–35; and Winkler, *William Randolph Hearst*, 43–50.

91. William Swanberg, *Citizen Hearst* (New York: Scribner, 1961), 33.

92. Ibid. Also see Proctor, *William Randolph Hearst: The Early Years 1863-1910*.

93. Mrs. Fremont Older, *William Randolph Hearst* (New York: D. Appleton, 1936); and Winkler, *William Randolph Hearst*.

94. Swanberg, *Citizen Hearst*, 90.

95. Clifford Krauss, "Remember Yellow Journalism," *New York Times*, February 15, 1998, Week in Review section, 3. A forthcoming book by Joseph Campbell, however, disputes this conventional wisdom and argues that the cable never was sent. See "Hot Type," *Chronicle of Higher Education*, September 10, 1999, A26.

96. Krauss, "Remember Yellow Journalism," 3.

97. Allan Keller, *The Spanish-American War: A Compact History* (New York: Hawthorn, 1969).

98. Jones, *Journalism in the United States*, 436.

99. Ibid., 438–440.

100. W. Weinfeld, "The Growth of Daily Newspaper Chains in the United States, 1923, 1926–1935," *Journalism Quarterly* 12 (1936): 357–380.

CHAPTER 4

The Objective Media

In 1928, a popular play *The Front Page* (later a movie) characterized reporters as tabloid creatures who ran from crime scene to crime scene, trying to get the most sensational tidbits to splash on the front page. Such reporters were not very knowledgeable, neither were they professionally trained nor terribly scrupulous in their pursuit of the news. If getting the story required bending the law or bribing a source, that was acceptable.

Typical of views about journalists around the turn of the century, this portrait disturbed the great editor Joseph Pulitzer. In the 1800s, many reporters worked their way up to editorial positions from entry-level typesetting jobs. The career path represented by these men — long apprenticeships followed by eventual promotions — was the norm. Even among editors, few people on nineteenth century newspapers held college degrees.[1]

As one of the leading journalists of his era, Pulitzer felt it was time to professionalize the industry. "What is everybody's business is nobody's business — except the journalist's. It is his by adoption," he once said. "But for his care almost every reform would be stillborn. He holds officials to their duty. . . . An able, disinterested, public-spirited press, with trained intelligence to know the right and courage to do it, can preserve that public virtue without which popular government is a sham and a mockery."[2]

It was this spirit that led Pulitzer to call for the creation of a journalism school (and to provide an endowment for one at Columbia University) that would professionalize the industry and instruct journalists in the same way that lawyers, doctors, and business leaders were trained. Writing in 1904, Pulitzer noted that "better results are obtained by starting with a systematic equipment in a professional school." The task of such a school, according to Pulitzer, was "to exalt principle, knowledge, culture, [and] to set up ideals." His hope was "that this college of journalism will raise the standard of the editorial profession."[3] With $2 million left in his will, Pulitzer founded Columbia's School of Journalism and created the Pulitzer Prize for outstanding investigative reporting. Columbia admitted its first class of aspiring journalists in 1912.

By 1940, more than six hundred colleges and universities had established some type of instruction devoted to journalism. The curricula emphasized courses in newswriting, editing, and reporting as well as classes on ethics and the history of journalism. The first journalism textbook, *The Practice of Journalism*, written by

Walter Williams and Frank Lee Martin, appeared in 1911 and helped train a new generation of journalists.[4] Other instructional materials soon followed.

From Pulitzer's initiative eventually evolved a powerful new journalistic stage known as the *objective media*. This era witnessed the professionalization of the media. Journalism schools instructed reporters in professional standards. News departments were separated from editorial and business departments. Reporters sought "the truth" and to evaluate politicians along clear standards. Although no journalist was thought to achieve these lofty ideals, outside observers appreciated the industry's effort to move beyond the partisan and commercial motivations of the previous century.

Over the course of several decades, this effort to professionalize the media produced a remarkable transformation in how the news was covered, how the public viewed the industry, and the amount of influence the media had. Control of the news moved from the powerful publishers and editors of the nineteenth century to professional reporters who argued that only they could fairly report the news, free of partisan bias and untainted by economic incentives.[5] Street reporters gained enormous power over news gathering, filed reports that were homogeneous in content and tone, and put themselves in a position where readers perceived them as having an extraordinarily high degree of trust and credibility. With high source credibility and a consistent message, people respected the media messenger, which gave journalists power to shape the country's agenda, frame the news, and define how problems got covered. This period became the peak years of American journalism, when reporters were celebrities and wielded considerable power over the political process.

THE RISE OF OBJECTIVITY

It is not an easy task to define objectivity. Although straightforward as a general notion, the idea has proven to be quite elusive in practice. In looking through writings on the subject, there are wide variations in the standards used to assess objectivity.[6] Some have emphasized baselines such as fairness, balance, and equity in coverage. Others have chosen to focus on ideas like neutrality and freedom from partisanship.

These difficulties notwithstanding, as soon as the industry began to move in the direction of objectivity, a press association provided a working definition to guide this newly emerging profession. Writing in 1923, the American Society of Newspaper Editors defined news objectivity as coverage that was "free from opinion or bias of any kind."[7] According to these "canons of journalism," the task of reporters was "freedom from all obligations except that of fidelity to the public interest." The society noted that "promotion of any private interest . . . is not compatible with honest journalism" and that "partisanship in editorial comment which knowingly departs from the truth does violence to the best spirit of American journalism."[8] In stating these principles, the professional association clearly sought to demarcate the goal of coverage in the twentieth century from the partisan and commercial tabloid tendencies of the nineteenth century.

Not only did the goal of coverage change, there is evidence during this time period that news coverage itself underwent a major transformation. Between 1865 and 1934, according to one study of local and wire reports, the percentage of all stories that were "objective" in nature (which the author defined as free from values) rose dramatically. One-third of all stories from 1865 to 1874 were objective, compared to one-half of stories from 1885 to 1894, two-thirds of those from 1905 to 1914, and four-fifths from 1925 to 1934.[9]

Many things contributed to this rise in objective coverage. The country was going through major changes at the beginning of the twentieth century. Education levels were rising; more people were graduating from high school. Whereas in 1880, just 2 percent of the adult population had graduated from high school, this number rose to 6 percent in 1900 and 16 percent in 1920. For the first time in the nation's history, mass education was seen as an explicit goal of public policymakers.

In addition, a variety of occupations such as law, medicine, and business were professionalizing. Trade associations successfully argued that it was important for practitioners of the craft of journalism to be trained in a field and undergo rigorous standards of evaluation. Such credentialing would include not just creating formal educational programs but sponsoring periodicals, holding conferences, and sending out newsletters to members. Thanks to the efforts of Pulitzer, Medill, and others, industry leaders began to upgrade the profession of journalism. Unhappy at the low status of the industry at the very time the field was earning greater and greater authority to report the news, journalism's titans felt it was important that reporters master their craft in the same way doctors and lawyers learned their fields.

Broader social trends were transforming America as well. Media historian Michael Schudson places the rise of the "ideal of objectivity" squarely in the context of various intellectual currents: a reaction against subjectivity, a fear that personal values would bias a story, and a fundamental faith in "the facts," illustrated by the famous signs that adorned newsrooms, exhorting reporters to attend to the "Who? What? Where? When? How?" of the news.[10] Throughout the country, there was a sense that things were changing and that news organizations needed to adapt as well. In an effort to increase personal accountability for news presentations, the *New York Times* started using "bylines" with reporters' names in the 1930s. This allowed readers to associate particular articles with individual authors. Shortly thereafter, the Associated Press adopted the same policy.[11]

New media such as radio and motion pictures were arising and attracted a mass audience. KDKA in Pittsburgh was the earliest commercial radio station; it provided detailed coverage of the 1920 presidential campaign. The intimacy radio provided listeners and the potential for news updates during the day represented a major advance over newspapers, which generally published only morning and evening editions. Radio listeners could hear about news events as they were happening. Within three years, more than six hundred stations were broadcasting live on the air. In 1924, there were 3 million radios in America; by 1930, this number had risen to nearly 14 million.[12]

The rise of radio reinforced the beliefs of journalistic leaders that it was desirable for reporters to adhere to professional norms. With the ability of radio signals to be transmitted over broad geographic areas and the formation of radio networks, national broadcasts became possible. This encouraged journalists to wring partisan and commercial excesses out of their programming in order to attract more listeners. If broadcasting was a goal as opposed to reaching a narrow audience, reporters needed to develop a more homogeneous and professional product that would appeal to different kinds of people.[13]

The necessity of professionalism had dramatic consequences for the way reporters did their job and trained to become journalists. For the first time, formal educational attainment was an important requirement for reporters. Journalists with college degrees became more common. Journalism schools flourished across the country, from Columbia University and the University of Missouri to Syracuse University and Northwestern University.

In these professional schools as well as in newsrooms themselves, reporters were taught to aim for objectivity. The goal was fairness, balance, and the search for the truth. During elections, the journalist's task was to report on all the serious candidates and provide equal time for the two major parties. If journalists provided fair and objective reports, voters could decide who was the best candidate.

It was not as if editors suddenly lost their historic mission of direct advocacy. Befitting the press's long tradition of commentary, journalists such as Walter Lippmann and I. F. Stone were as opinionated as ever. But editorial departments, encompassing columns, commentaries, and opinion pieces, were separated from news departments. It was acceptable to be opinionated, but exhortations should appear on the editorial page, not the front page.

Between elections, the journalist's goal was to evaluate government performance. Were leaders performing well? Reporters began to investigate how leaders were doing their jobs, from the muckrakers of the Progressive Era to the investigative reporters of the 1960s and 1970s. As evidenced by several cases, this era was marked by detailed exposés on leadership and governmental activities that helped raise the credibility of the media profession. In the movement from powerful editors, such as Greeley in the nineteenth century, to investigative reporters, such as Tarbell, Steffens, Halberstam, and Woodward and Bernstein in the twentieth century, journalists defined a new role for themselves that liberated them not only from outside parties and advertisers, but even from the owners and publishers within the communications industry.

THE PROGRESSIVE MOVEMENT

One of the earliest examples of changing journalistic mores came during the Progressive Era. Conditions in America around the turn of the twentieth century were intolerable for many people. In the days before the graduated income tax was approved in 1913, there were enormous variations in wealth between the rich and poor. "Robber barons" took advantage of weak political leadership to build monopolies in oil, steel, and railroads, among other industries. They then used their market muscle to further enrich themselves and beat back the competition.

Some journalists, known as "muckrakers," took exception to these conditions and wrote powerful stories condemning this private enrichment at the expense of the general public. The muckrakers publicized deplorable housing and sanitation in American cities and cited the need for governmental action. One notable newspaper publisher, Edward Scripps, explained his guiding personal principle was "to make it harder for the rich to grow richer and easier for the poor to keep from growing poorer."[14] In conjunction with editors such as Adolph Ochs, who bought the *New York Times* in 1896, these men sought to raise their industry to new heights.

Taking advantage of popular discontent, the social force known as the Progressive movement sought fundamental change in the political process. Its adherents believed that partisan politics should be removed from the governmental process and that government administration should be professionalized. Managers both in industry and government should be trained to make professional decisions based on the facts, not on personal beliefs or political connections.

This belief in the virtue of professionalism had as many consequences for politics and business as journalism. In the political realm, the last part of the nineteenth century and the early twentieth century saw many academics who felt public policy should be left to public administrators who could make rational decisions "in the public interest" away from partisan political processes.[15] The party machines that dominated the turn-of-the-century political system were corrupt and interfered with effective government.

In the business world, scientific management principles advocated by Frederick Taylor were becoming popular. As in the emerging field of public administration, business leaders sought to run their enterprises in more professional ways. Professional managers sought to remove redundancy and duplication from factory floors and run their businesses in a scientific manner. Time management studies provided data on how best to maximize the efficiency of business operations.[16]

Given these broad trends sweeping America, it was little surprise that similar sentiments arose about reform in the journalism field. During this period, newspapers sought to cover political developments according to more objective criteria. The goal was not commercial exploitation of the news market, as had been the case at the turn of the century, but a search for the truth, however ambiguously that notion was defined.

The rise of newspaper syndicates and wire services encouraged more homogeneous news coverage. Because news stories no longer served a single locality but rather ran in newspapers all across the country, it was crucial to squeeze personal biases and local idiosyncrasies out of coverage. Pressures dictated by the need to sell ads mandated more neutral and generic news stories. Only by running relatively impartial stories could newspapers attract ads from disparate business organizations across the country, each of which had their own political viewpoints.[17]

The immediate goal of the Progressives was to find out why American politics was functioning so poorly. Corruption had become widespread and many ordinary citizens felt their views were not being heard by elected representatives. The villains, in the eyes of many, were the industrial robber barons and the corrupt politicians they bought to do their bidding. According to the Bureau of the Census in

1914, "one-eighth of American businesses employed more than three-fourths of the wage earners and produced four-fifths of the manufactured products."[18] Economic power was particularly concentrated in the industries of steel, oil, copper, sugar, tobacco, and shipping.

To make matters worse, rich men used bribes and payoffs to leading politicians to exert their will on the political system. Graft was prevalent in large cities around the country and in state capitals and Washington, D.C. The tide of corruption was so widespread that grassroots organizations were springing up to demand change, both in the political and economic spheres.

Journalists at a variety of papers and magazines wrote exposés criticizing the concentration of wealth and the political abuses which followed from it. One of the most famous investigative pieces was Ida Tarbell's "History of the Standard Oil Company." This series of articles was published in *McClure's* magazine around the turn of the century. Each report documented the unfair business practices used by John D. Rockefeller to build up his oil company, and helped pave the way for the eventual government breakup of Standard Oil's monopoly into competing corporations.[19]

Lincoln Steffens meanwhile focused on political corruption. In a series of articles about particular cities and states across America, he documented cash payoffs and corrupt political machines. Few cities were immune to the power of private money. Party bosses used bribes to enrich themselves, extend their political reach, and dominate the electoral process.[20]

Famed editor William Randolph Hearst himself launched a personal crusade for reform. Built around the ideas of a personal income tax and the "destruction of criminal trusts," he sought fundamental changes in how the political process worked. Hearst later would run unsuccessfully for public office, in a vain effort to reform the political system.

Although the Progressive movement eventually flamed out as a political force following World War I, many of its ideals would shape American politics and journalism for decades to come. A number of political reforms it engendered endured — from primaries, initiative, and referenda to a professionalized civil service. New ethics rules were adopted that weakened political parties. And in the journalism field, reporters gained new autonomy to search for truth as defined by their own professional standards. It was a dramatic example of how a social movement lost the short-term battle, but transformed American society.

WORLD WAR II

World War II presented many challenges for the American media, but afforded journalists who were far away from home offices great power over the presentation of the news. Over sixty countries were involved in the war and more than 75 million troops served in the various militias of these nations. Of these, 15 million were killed and 25 million wounded. One estimate put the financial cost of the war at $1 trillion, with property damage running around $230 billion. Based just on sheer numbers, it was "the biggest and most tragic war in the history of the world."[21]

More than 2,600 American correspondents were sent to cover the war at various locations around the globe, a sign of the vast public interest in the conflict. Every leading newspaper had reporters on the front lines to describe the war, as did radio networks and the wire services. According to U.S. War Department listings, 1,800 correspondents were accredited with the Army and 800 accompanied the Navy.[22] Coverage was designed to ensure that people back home had detailed information about what was happening.

Much as they had done during the Civil War, journalists had to walk a fine line between covering the war and not endangering national security. Propaganda flowed freely from both sides. German propaganda ministry officials, for example, went to great lengths in order to convey their point of view to foreign correspondents. Describing the scene of large German press conferences held in a Berlin theater, CBS Radio correspondent William Shirer said, "It seats about 200 people in very comfortable upholstered chairs, facing the stage where the officials sit. On the stage, as a sort of backdrop, there is a huge illuminated map where the High Command boys used to try to tell us what it was all about. Goebbels himself occasionally dropped in."[23]

As they had in the nineteenth century with the telegraph, journalists took advantage of the then relatively new technology of radio to file the most noteworthy reports from the front lines. The percentage of American households with a radio rose from 1 percent in 1922 to 46 percent in 1930, to 82 percent in 1940, and 94 percent in 1950 (see Appendix Figure A.2). Radio broadcasts turned individual reporters into international celebrities and lay the groundwork for even higher source credibility for the profession.

Initially, newspaper editors worried about the threat posed by radio and fought to protect their readership. The new medium provided an immediacy that was lacking with newspapers and wire service reports, and radio stations could broadcast information directly to listeners without going through newspapers. At first, fearing the power of radio, newspapers and wire services refused to run radio program listings in their papers. Radio stations were forced to purchase ads to get program listings into print outlets. Eventually, an accommodation was reached and newspaper titans figured out their best strategy was to buy radio stations in order to combine and expand their communications reach.[24]

The war produced new media celebrities such as Shirer, Edward R. Murrow, Bill Henry, Arthur Mann, and William Kerker. During 1940, Murrow reported to CBS listeners the horrors of the Battle of Britain, the massive German air raids on London. As one writer put it, "Murrow's quiet but compelling voice brought images of a bomb-torn and burning London that did much to awaken the still-neutral United States to the nature of the war."[25]

After the United States entered the war, correspondents blanketed the front lines in Europe, Africa, and Asia to satisfy endless public curiosity about the progress of the war. Using mobile units and tape recorders, journalists provided coverage that was, according to one observer, "the best and fullest the world had ever seen."[26]

Even cartoonists got into the act of reporting the war. Bill Mauldin drew a one-panel cartoon called "Up Front" for newspaper readers throughout the war. His cartoon developed the quintessential image of the American infantryman:

"craggy-faced, unshaven, disheveled, [and] sardonic." Mauldin felt that "if anybody fought an idealistic war, it was the United States."[27]

The national government actively tried to influence the course of war coverage. Among the most important federal agencies during the war was the Office of War Information, which staffed the Voice of America radio broadcasts. According to John Houseman, a producer with VOA, by summer of 1941, the country had "750 broadcasts a week," many in the form of short-wave broadcasts into enemy territory.[28]

Following the war, correspondents and photographers felt a sense of euphoria over what they had accomplished. Photojournalist Walter Rosenblum described the experience this way: "To see fascism defeated, nothing better could have happened to a human being. You felt you were doing something worthwhile. You felt you were an actor in a tremendous drama that was unfolding. It was the most important moment in my life."[29]

THE EXPANSION OF MASS EDUCATION FOLLOWING THE WAR

With the task of reintegrating American troops into national life, the federal government decided to finance a dramatic expansion in public education. Through the GI bill and direct grants to secondary and higher education, there was a significant increase in numbers of people who went to college. Indeed, the post–World War II period was the first one in American history to see college education move beyond elite groups to mass society.

This societal change had profound consequences for the media. The percentage of journalists who were college graduates rose rapidly after the war. In 1936, according to a systematic survey of the Washington press corps, Leo Rosten found that 51 percent of reporters had a college degree.[30] By 1961, when William Rivers followed up with a new survey of the Washington press corps, 81 percent had graduated from college. And in 1978, a survey of Washington journalists by Steven Hess revealed that 93 percent were college graduates.[31] In addition, nearly half of these correspondents in Hess's study had done some graduate work and one-third actually held graduate degrees.[32]

It was not just Washington correspondents who were highly educated. In a 1979 and 1980 study of 238 journalists who worked at the country's most influential media outlets, Robert Lichter found that 93 percent were college graduates and a majority had attended graduate or professional programs too. He concluded that journalists were "one of the best-educated groups in America."[33] Drawing on this and other findings, Lichter wrote a book describing journalists as the "media elite" and "America's new powerbrokers." In contrast to the yellow journalism tendencies that characterized nineteenth-century journalists, twentieth-century reporters were "highly educated, well-paid professionals."[34]

Aside from changes in educational levels, the industry itself was undergoing a dramatic revolution. First was a major change in the newspaper business. The percentage of U.S. cities having two or more newspapers dropped from 61 percent in 1923 to 2 percent in 1992.[35] One-newspaper towns, previously the exception rather than the rule, became the norm in most major American cities after the war. This gave newspapers a virtual monopoly over the print industry.

In addition, television emerged as a major new challenger to newspapers and radio. Combining the features of radio with the visual technology of moving pictures, television represented a potent combination for the news industry — one that attracted a mass audience of huge proportions.

Television had several qualities that viewers appreciated. First was its immediacy and ability to convey events as they were happening. Television cameras could give viewers a "you are there" perspective on contemporary events. For the first time in recorded history, viewers had front-row seats at the most important events as they were occurring. Second was the visual dimension that played to peoples' fascination with pictures. Unlike either radio or newspapers, television could show viewers what was happening, not merely describe what had taken place. Perhaps nowhere were these two qualities more evident than in television's coverage of political events, such as nominating conventions, speeches, and rallies. Everything from political campaigns to presidential press conferences could be shown live to a national audience, providing viewers with a sense that they were witnessing events first-hand.

As radio and television developed, it was not clear whether they should be regulated in the manner of other public utilities, such as railroads, telephones, and electric companies, or be left to the marketplace and financed by commercial advertisers. In European countries, for example, control of these news communications technologies was considered so important to democracy that systems of state-run channels with virtually no advertising developed. After extensive campaigns by American radio and then television stations for a market model, these media developed as commercial properties financed in large part by advertising, but subject to government oversight, professional norms, and industry self-regulation. This approach was ratified by the landmark 1934 Communications Act, which formalized how these new technologies would develop.

Eventually, the particular virtues of television would converge in a way that enhanced its audience size and the credibility of the media industry as a whole. In 1959, during the infancy of television, newspapers still were cited by Americans as their primary source of news. Fifty-seven percent named newspapers, followed by 51 percent for television, 34 percent for radio, 8 percent who named magazines, and 4 percent who said other people. However, starting in 1963 and continuing thereafter, more people named television as their most important news source. That year, 55 percent named television, compared to 53 percent for newspapers and 39 percent for radio. By 1986, two-thirds of Americans (66 percent) were naming television as their top news source, 36 percent were saying newspapers, and 14 percent were citing radio.[36]

The rise of television viewership had profound consequences for political coverage and the power of reporters in the political system. Almost as soon as television emerged, there was a drop-off in overall newspaper circulation levels. Newspaper subscriptions decreased from its historical high point of 37 percent in 1947 to 26 percent in 1986 (see Appendix Figure A.1).

With the decline of newspapers and concurrent rise of television, public officials and candidates for office increasingly started to play to the television cameras in order to reach a mass audience. In 1950, only 9 percent of American homes had television sets. But as this number rose to 88 percent in 1960 and 95 percent in 1970 (see Appendix Figure A.2), television reporters became major political gatekeepers. Previously more attuned to party leaders, candidates sought to cultivate journalists in an effort to shape public perceptions of themselves. Through both free and paid media, television began to dominate American politics. The 1952 presidential campaign was the first one to feature television advertisements for candidates. Although primitive, often featuring stilted testimonials or clips from actual news conferences, these ads marked a new communications era emerging in the United States. By the 1980s, television advertising comprised nearly 60 percent of the overall presidential campaign budget for candidates.[37] The 1960 presidential election was the first one in which televised debates were thought to have been decisive in the fortunes of the candidates. The contrast between the youthful John F. Kennedy and perspiring Richard Nixon accentuated the power of television and gave us campaigns that would come to be known as "mass media elections."

THE EVALUATION OF SCHOLARS

The outpouring of stories by professionally trained journalists during the war and postwar periods created materials for which academic scholars could assess the quality of news coverage. With the goal of objectivity at center stage, academic researchers set about the task of evaluating press coverage. How well were the media doing based on their own aims? Was the coverage fair and objective? Scholars spent hours counting stories for different candidates at election time and evaluating the tone of coverage.

One of the earliest studies was of the 1940 presidential campaign in Erie County, Ohio, between Franklin Roosevelt and Wendell Wilkie. Using an analysis of newspapers, magazines, and radio, Paul Lazarsfeld, Bernard Berelson, and Hazel Gaudet found that the press fell short of objective standards. Writing about newspapers in the area, the authors concluded, "If the papers were strictly objective in their reporting, the readers of all three would have been given roughly the same picture of the campaign, in terms of over-all partisanship. Actually, however, they were not." After following the newspaper coverage, the authors found that "the Republican paper managed to stress the Republican side, the Democratic paper managed to stress the Democratic side, and the neutral paper . . . managed to strike a balance between the other two."[38]

A study by Thomas Patterson and Robert McClure entitled *The Unseeing Eye* examined whether Richard Nixon and George McGovern got the same type of network television coverage in the 1972 election. The scholars found the coverage was more extensive for Nixon (eighty-four minutes of policy coverage and twenty-eight minutes devoted to his personal qualities) than McGovern (twenty-four minutes of policy coverage and sixteen minutes devoted to his personal qualities). But there were obvious reasons for the differences: Nixon was an incumbent president who was coasting toward an easy reelection while McGovern was seen as an out-of-the-mainstream challenger with little chance of winning.[39] The inequity was not unexpected given the relative political positions of the two men.

Michael Robinson and Margaret Sheehan performed the same task for the 1980 election in a book titled *Over the Wire and on TV.* Examining campaign coverage by CBS and UPI, the authors studied the objectivity of the reporting, which they defined as neutral coverage without a point of view. Their conclusion was that "CBS and UPI came close to being objective."[40] On the dimension of balance, media outlets provided "equal time for equal actors," meaning that serious candidates got treated fairly in the amount of coverage. However, reporters got to define who was "serious."[41]

The results of these and other academic studies often chided reporters for failing to be as fair, complete, and balanced as professional standards warranted. By setting up such a clear benchmark for evaluation, professional journalists almost invariably failed to measure up to their own industry standards. But most academics appreciated the effort by journalists to seek objectivity, even if it fell short of their goal. Implicit in most research projects was the normative ideal that objectivity was far preferable to the partisanship and yellow journalism practiced by many nineteenth-century editors and reporters.

The only real exception to this generally sympathetic evaluation of twentieth-century media was television, an industry that arose in the 1940s and then became very prominent in the 1950s and 1960s. Scholars almost always judged television more harshly than newspapers. Television coverage repeatedly was described as trivial and superficial compared to that of newspapers, which was seen as more complete and in-depth. With its emphasis on pictures, not substantive content, observers felt television weakened citizens' ability to be informed about public matters.

For example, C. Richard Hofstetter undertook a detailed critique of network television coverage in the 1972 presidential campaign. Using the definition of bias as "partiality in news programming," Hofstetter found there was little evidence of partisan political biases, but numerous cases of structural and situational biases. Nixon, for example, received more favorable coverage than McGovern from the three television networks.[42]

Todd Gitlin meanwhile condemned CBS stereotyping of the 1960s protest movement known as the New Left. Due to the twenty-two-and-a-half-minute length restriction of the television evening news, television simplified stories to the point that viewers received cartoon-like summaries of Students for a Democratic Society in 1965. In Gitlin's mind, "the imperative of finding 'good pictures' (usually vivid illustrations) adds to the premium on simplification."[43]

Others defined television as a cultural "wasteland," devoid of in-depth coverage and superficial in its treatment of substantive issues. Rather than being a new medium that would use visual technology to take citizens to a higher plane of political understanding, television was thought to play to society's worst instincts. In the eyes of critics, coverage typically emphasized vacuous programming and uninformative reporting.[44]

Despite such condemnations of television, the media during this period was seen as much more satisfactory in their general approach to news gathering (i.e., the effort at objectivity) than had been true in the nineteenth century. The idea of "progress" was engraved indelibly into people's consciousness, and many observers concluded that media professionalization and objectivity were clear improvements over what had existed in earlier times.

Taking advantage of their new professional training and a growing ability to hold political leaders accountable, reporters achieved two notable successes in their coverage of Vietnam and Watergate that would impress both academic scholars and the general public. Prominent CBS commentator Eric Severeid would conclude in regard to the mid-twentieth century that "this has been the best generation of all in which to have lived as a journalist in this country."[45]

THE CASE OF VIETNAM

Vietnam started as a determined effort by Presidents John F. Kennedy and, following his 1963 assassination, Lyndon Johnson to stem the tide of communism around the world. The object of the war was a small Asian country at the southern edge of China. Financed by Russia and the Chinese, the North Vietnamese were attempting to expel the South Vietnamese government first supported by France and then the United States from the Indochinese peninsula and reunite their country. It was a test of America's bipartisan postwar foreign policy doctrine of containment, by which communism was to be stopped from expanding anywhere around the world.

In the early days of the war, America seemed to be very successful militarily. Although Johnson had campaigned in 1964 as a peace candidate, he dramatically expanded the war effort after the election. Within a year, the United States had nearly 500,000 troops in Vietnam and was bombing major targets throughout the small country. Daily body counts revealed an increasing number of Viet Cong troop deaths. In the face of superior manpower and military equipment from the United States, it seemed only a matter of time before America triumphed. Saturation bombing runs by the Air Force appeared to be knocking huge holes in North Vietnamese defenses.

But perplexing disparities started to arise. Though body counts climbed higher and higher, there seemed to be more North Vietnamese troops than ever. Before long, both television and newspaper journalists were investigating the military and finding clear evidence of deceit, both in terms of body counts and general war conduct.

For journalists, Vietnam would not prove to be a rerun of World War II. In that war, the enemy had been clear-cut and the moral stakes unambiguous. The very future of civilized society was in doubt. Vietnam was not anything like that. The enemy was amorphous and moral stakes were textured in various hues. The United States itself grew quite divided as policymakers and ordinary people debated the proper course of action. In the end, the Vietnam war produced a defeat for the United States but a triumph for investigative journalists.

The conflict brought a host of bright, young, and aggressive reporters such as Neil Sheehan, Morley Safer, and Peter Arnett to the forefront. These and other journalists were dedicated to pursuing the truth wherever it led, even at the risk of life and limb. A number were killed in the line of journalistic duty. Others lived and earned Pulitzer Prizes for their courageous efforts.

One of the best newspaper reporters was David Halberstam, who later wrote a book about the Vietnam experience called *The Best and the Brightest*. In that book, Halberstam showed how the press looked beyond the formal statements of government officials. In so doing, reporters often discovered a picture that was far different than what the American public was being told.[46] Statistics were deliberately fabricated to preserve the facade that America was winning the war. Month after month, reporters were informed that things were going well and North Vietnamese resistance was slowing down.

The problem with these official statements of reassurance, both from the military and President Johnson, was that they simply were not true. Journalists at the front lines and in Washington uncovered persuasive evidence that the war was not going well and that government officials were deceiving the public. More importantly, taking advantage of the visual power of television, the war was broadcast live into American homes. Vietnam was the first real television war, and the pictures of body bags containing slain U.S. troops were quite grim.

One of the most vivid illustrations of this occurred in 1968 during the so-called Tet offensive. After months of being told the country was winning the war, Americans were shocked when North Vietnam launched a major offensive that succeeded in overrunning several major cities in the South, including the capital city of Saigon.[47] This development provided the clearest possible evidence to the American public that what investigative reporters were saying was correct — the war was in danger of being lost.

This event had a major impact on press coverage. After the Tet offensive, the number of news stories with negative references to U.S. troop morale rose from 0 to 14.5 while positive references dropped from 4.0 to 2.5.[48] More and more domestic critics of the war were getting air time, arguing that the war was going nowhere and that innocent lives were being lost for no good reason. The consistent message Americans were starting to get about the war was that it was not going well and that top government officials were lying.

To make matters even worse, investigative reporters found that American troops were committing atrocities on the Vietnamese civilian population. Unable at times to distinguish Viet Cong troops from innocent civilians, military forces were accused of misconduct. In one particularly gruesome and reprehensible display, troops at My Lai massacred unarmed women and children. Reporting on this event and the subsequent military cover-up earned Seymour Hersh a Pulitzer Prize for investigative coverage.[49]

As the widespread deceit on the part of government officials became apparent, time and time again reporters exposed the president and military establishment as not being truthful. In conjunction with the rising death toll, this coverage undermined public support for war and led to an increase in public cynicism about the government.[50] In 1968, public pressures and the prospects of a divisive primary forced Johnson to announce he would not seek reelection. In the end, the Vietnam War strengthened the role of journalists in communicating news to the public and encouraged future reporters to probe even further into the conduct of government policy.

THE WATERGATE SCANDAL

It wasn't just war that demonstrated the newfound influence of reporters. On the heels of Vietnam came a domestic crisis that would propel American journalists to even greater heights of personal credibility. Watergate, as the scandal was known, started as a minor burglary of Democratic National Committee (DNC) headquarters in Washington, D.C. Amateurish burglars at the Watergate complex put masking tape around the door frame to keep it unlocked. Walking around the office building late that night, a security guard on patrol noticed the tape, called the police, and in so doing altered the course of American history. The D.C. metropolitan police arrived, found five men inside DNC headquarters, and placed them under arrest. Eventually, these arrests would unseat a president and make heroes of investigative reporters.

At first, Nixon's press secretary Ron Ziegler downplayed the caper as a "third-rate burglary." White House insiders questioned the significance of the case. The burglars were acting independently and had no connection with the president, they argued.

But gradually, these claims proved more and more hollow. Indeed, the truth was earth-shattering. In 1971, a frantic Nixon, who was worried about his upcoming election, had instigated a secret "plumbers" unit designed to investigate embarrassing leaks from the executive branch. With lots of secret money to spend, this unit soon was expanded to a "dirty tricks" campaign which undermined the efforts of emerging Democratic opponents such as Ted Kennedy and Edmund Muskie.

Rather than being an isolated event, the break-in was part of a systematic plan of surveillance and disruption targeted at Democratic opponents by the Nixon White House — a brazen effort to damage the political opposition and secure Nixon's own reelection in 1972. The dirty tricks campaign was about as close as America ever came to unmaking one of our most cherished democratic processes — competitive presidential elections.

For months, as journalists, members of Congress, and judges circled around in a vain attempt to uncover the truth, the White House stonewalled the investigation, citing both national security and executive privilege arguments. But two persistent reporters from the *Washington Post*, Bob Woodward and Carl Bernstein, had a secret source named Deep Throat (the name of a porn movie of that era) who guided them through the maze of Watergate. Holding secret meetings in parking garages and engaging in furtive phone calls, these reporters helped to break open the scandal, revealing its reach to the highest level of government.

The public initially was reluctant to believe the accusations. Polls during the 1972 campaign, for example, found that many people were dubious of the charges. Whereas 29 percent felt Nixon needed to "answer and explain these charges,"

51 percent believed McGovern's criticisms were "simply a political charge by a desperate politician who is behind."[51]

It would take months to prove, but a secret White House taping system installed by Nixon in the Oval Office eventually undid the president. After denying any personal culpability for the original break-in or the ensuing cover-up, the Nixon tapes proved the truth was much more damaging. Not only was the president implicated in the scandal, the tapes demonstrated that the planning reached into the very top level of the Nixon administration. Between September 1972 and April 1973 when press reports started to intensify, public awareness of Watergate rose from 52 to 83 percent.[52]

Saturation coverage of the Watergate tapes saga, congressional hearings, court proceedings, and the weakening domestic economy eventually had an impact on the general public. In August 1974, after the public revelation of tapes containing damaging statements, the president was forced to resign. Nearly three-quarters of Americans had come to believe in Nixon's "complicity" in Watergate.[53]

Woodward and Bernstein became national heroes for exposing misdeeds at the highest level of the government. Eventually, a book they wrote entitled *All the President's Men* would become a national best-seller and be turned into a popular movie, in which Dustin Hoffman played Bernstein and Robert Redford portrayed Woodward.[54]

In uncovering massive government deceit in Vietnam and Watergate, this era of objective media became the glory period of American journalism. Reporters presented themselves as fighting for truth and justice, and citizens responded by according tremendous credibility to the media messengers. Journalism became a prestigious profession, often accompanied by lucrative salaries. News coverage was homogeneous in content and tone, as reporters emphasized a professional product based on industry norms.

In 1972, it was documented that more people trusted CBS anchor Walter Cronkite than the president of the United States. According to an Oliver Quayle poll of 8,780 people conducted in eighteen states, Cronkite headed the list of major public figures. Whereas 73 percent indicated they trusted Cronkite, 67 percent said they trusted the average senator, 59 percent the average governor, 57 percent President Nixon, 54 percent New York City Mayor John Lindsay, and 50 percent Vice President Spiro Agnew.[55]

THE PINNACLE OF SOURCE CREDIBILITY

With several demonstrated examples of successful investigate reporting, public confidence in the American press was quite high in the 1960s and 1970s. Compared to European countries, American reporters were seen as more fair and credible by its citizens, compared to the views of the English about the British press and Germans about their media. Whereas 69 percent of Americans expressed great confidence in their media, only 41 percent of Germans did, and 38 percent of British did.[56]

Television also came to be seen by Americans as more "believable" than newspapers or radio. In 1959, more people (32 percent) claimed newspapers were believable than television (29 percent). However, by 1968 in the middle of the Vietnam conflict, 44 percent said television was believable compared to 21 percent who felt that way about newspapers. And in 1974, during the heart of Watergate, 51 percent felt television was believable and only 20 percent felt that about newspapers.[57] Television again had trumped its media rivals for the heart and soul of the American public.

These numbers provide vivid evidence of how the media during this period came to exercise considerable political power, employing a combination of homogeneous coverage, believable reporters, and visual power. Between Vietnam and Watergate, negative press coverage along with actual events played a role in toppling two consecutive presidents.

Persuasion studies often have shown that the degree of public persuasion is dependent on the "credibility" of the source and receiving a consistent message from different outlets. People with high credibility are more believable than those who are not, and viewers exposed to a homogeneous, professional news product hear the same message repeated over and over again from respected sources.

In an era when people viewed politicians and businesspeople as not very trustworthy, American journalists accumulated great power because of their reputation for honesty and believability. In the 1950s, two-thirds of Americans trusted the government in Washington to do what was right. By the 1970s, however, two-thirds of Americans said they did not trust the federal government to do what was right.[58]

Both Richard Nixon and his vice president Spiro Agnew criticized the media for a liberal bias. Claiming that reporters were systematically opposed to the Republican administration, Agnew in particular gave a blistering series of speeches on this topic. Journalists and administration critics were nothing more than "pointy-headed intellectuals" and "nattering nabobs of negativism," Agnew complained. By engaging in "instant analysis" following every presidential speech, the press sought to undercut Nixon's authority. While in the long run, such complaints about press bias would sink in with the public, in the short run, they were to no avail.

The high source credibility and homogeneous product of American journalists had major ramifications for the political process. The media's extensive degree of moral authority grew to the point that elections were described as mass media elections and the governing process required formal strategies for news management.[59] No other outside participants in the political system accumulated as much influence as journalists.

In marked contrast to the previous century when an angry Henry Clay threw an Associated Press reporter out of his speech for failing to request permission to cover the event, public officials would now go to great lengths to solicit media coverage of their activities. Without the press, political success was not thought possible. More than one successful politician groveled before the Washington press corps in order to win favorable coverage for a political cause. The 1960s and 1970s represented the pinnacle of power for American journalists. Not only had they carved out professional autonomy for themselves within media organizations, they had persuaded the public that reporters were the best-equipped to provide fair, balanced, and informative news coverage.

NOTES

1. Robert W. Jones, *Journalism in the United States* (New York: E. P. Dutton, 1947), 503.

2. George Henry Payne, *History of Journalism in the United States* (New York: D. Appleton, 1920), 368–369.

3. Jones, *Journalism in the United States*, 512–513.

4. Walter Williams and Frank Martin, *The Practice of Journalism* (Columbia, Mo.: E. W. Stephens, 1911). Also see Jones, *Journalism in the United States*, 509.

5. My thanks to John Zaller for suggesting this argument to me. Also see Michael McGerr, *The Decline of Popular Politics: The American North, 1865–1928* (New York: Oxford University Press, 1986), especially chapter 5.

6. Michael Robinson and Margaret Sheehan, *Over the Wire and on TV: CBS and UPI in Campaign '80* (New York: Russell Sage Foundation, 1980), see chapter 3 for a review. Also see Gaye Tuchman, "Objectivity as Strategic Ritual: An Examination of Newsmen's Notions of Objectivity," *American Journal of Sociology* 77 (1972): 660–679; and Rodger Streitmatter, *Mightier Than the Sword: How the News Media Have Shaped American History* (Boulder, Colo.: Westview Press, 1997).

7. Robinson and Sheehan, *Over the Wire and On TV*, 39.

8. The text of the Society's Canons are listed in the appendix to Jones, *Journalism in the United States*, 701–703.

9. Cited in Michael Emery and Edwin Emery, *The Press and America: An Interpretive History of the Mass Media*, 8th ed. (Boston: Allyn & Bacon, 1996), 185. The original study is by Harlan Stensaas, "The Rise of Objectivity in U.S. Daily Newspapers, 1865–1934" (Ph.D. dissertation, University of Southern Mississippi, 1986).

10. Michael Schudson, *Discovering the News: A Social History of American Newspapers* (New York: Basic Books, 1978), 144–159.

11. Ibid., 145.

12. Kathleen Hall Jamieson, *Packaging the Presidency*, 3rd ed. (New York: Oxford University Press, 1996), 19.

13. Susan Smulyan, *Selling Radio: The Commercialization of American Broadcasting, 1920–1934* (Washington, D.C.: Smithsonian Institution Press, 1994), chapters 1 and 2.

14. Emery and Emery, *The Press and America*, 211.

15. Luther Gulick, *The National Institute of Public Administration: A Progressive Report* (New York: National Institute of Public Administration, 1928).

16. Frederick Taylor, *The Principles of Scientific Management* (New York: Harper & Brothers, 1911).

17. Gerald J. Baldasty, *The Commercialization of News in the Nineteenth Century* (Madison: University of Wisconsin Press, 1992).

18. Emery and Emery, *The Press and America*, 212.

19. Ibid., 219, 228.

20. Lincoln Steffens, *The Shame of the Cities* (New York: McClure, Phillips, 1904).

21. Robert Desmond, *Tides of War: World News Reporting 1940–1945* (Iowa City: University of Iowa Press, 1984), 448.

22. Ibid., 448–449.

23. Matthew Gordon, *News Is a Weapon* (New York: Knopf, 1942), 145–146.

24. Susan Douglas, *Inventing American Broadcasting, 1899–1922* (Baltimore: Johns Hopkins University Press, 1987); Susan Smulyan, *Selling Radio*; and Gwenyth Jackaway, *Media at War: Radio's Challenge to the Newspapers, 1924–1939* (Westport, Conn.: Praeger, 1995).

25. Emery and Emery, *The Press and America*, 344.

26. Ibid., 349.

27. Studs Terkel, *"The Good War": An Oral History of World War Two* (New York: Ballantine, 1984), 358, 361.

28. Ibid., 349.

29. Ibid., 382.

30. Leo C. Rosten, *The Washington Correspondents* (New York: Harcourt Brace, 1937), 159; also see reprint edition, Arno Press, 1974.

31. Stephen Hess, *The Washington Reporters* (Washington, D.C.: Brookings Institution, 1981), 83. The original Rivers study is discussed in William L. Rivers, "The Correspondents after Twenty-five Years," *Columbia Journalism Review* 1 (Spring 1962).

32. Stephen Hess, *The Washington Reporters*, 45.

33. S. Robert Lichter, Stanley Rothman, and Linda S. Lichter, *The Media Elite: America's New Power-brokers* (Bethesda, Md.: Adler & Adler, 1986), 21–22.

34. Ibid., 21.

35. Shirley Biagi, *Media/Impact*, 2nd ed. (Belmont, Calif.: Wadworth, 1994), 65.

36. Harold Stanley and Richard Niemi, *Vital Statistics on American Politics*, 3rd ed. (Washington, D.C.: Congressional Quarterly Press, 1992), 75.

37. Darrell M. West, *Air Wars: Television Advertising and Election Campaigns, 1952–1996*, 2nd ed. (Washington, D.C.: Congressional Quarterly Press, 1997), chapter 2.

38. Paul Lazarsfeld, Bernard Berelson, and Hazel Gaudet, *The People's Choice* (New York: Duell, Sloan, & Pearce, 1944), 113.

39. Thomas Patterson, and Robert McClure, *The Unseeing Eye* (New York: Putnam, 1976), chapter 1.

40. Robinson and Sheehan, *Over the Wire and on TV,* 57.

41. Ibid., chapter 4.

42. C. Richard Hofstetter, *Bias in the News: Network Television Coverage of the 1972 Election Campaign* (Columbus: Ohio State University Press, 1976), 199.

43. Todd Gitlin, *The Whole World Is Watching: Mass Media in the Making and Unmaking of the New Left* (Berkeley: University of California Press, 1980), 266. Also see Edward J. Epstein, *News from Nowhere: Television and the News* (New York: Random House, 1973), 17–18.

44. Newton Minow, *Abandoned in the Wasteland* (New York: Hill and Wang, 1995).

45. Lichter, Rothman, and Lichter, *The Media Elite: America's New Powerbrokers*, 1.

46. David Halberstam, *The Best and the Brightest* (Greenwich, Conn.: Fawcett Publications, 1969). Also see Kathleen Turner, *Lyndon Johnson's Dual War: Vietnam and the Press* (Chicago: University of Chicago Press, 1985).

47. Peter Braestrup, *Big Story: How the American Press and Television Reported and Interpreted the Crisis of Tet 1968 in Vietnam and Washington* (Boulder, Colo.: Westview Press, 1977).

48. Daniel Hallin, *The Uncensored War* (New York: Oxford University Press, 1986), 180.

49. Seymour Hersh, *Chemical and Biological Warfare* (Indianapolis: Bobbs-Merrill, 1968).

50. Michael Robinson, "Public Affairs Television and the Growth of Political Malaise: The Case of the Selling of the Pentagon," *American Political Science Review* 70 (June, 1976): 409–432.

51. Gladys Lang and Kurt Lang, *The Battle for Public Opinion: The President, the Press, and the Polls during Watergate* (New York: Columbia University Press, 1983), 43.

52. Ibid., 44.

53. Ibid., 310.

54. Carl Bernstein and Bob Woodward, *All the President's Men* (New York: Simon & Schuster, 1974).

55. Poll cited in the *New York Times*, May 25, 1972, 91. Also see Jerry Carroll, "Whom Do We Trust? Why, Walter Cronkite, Of Course," *Orange County Register,* November 1, 1990, J10. For other polls, see Walter Cronkite, *A Reporter's Life* (New York: Knopf, 1996), 209.

56. Laurence Parisot, "Attitudes about the Media: A Five-Country Comparison," *Public Opinion* 10 (Jan./Feb. 1988): 18–19, 60.

57. Stanley and Niemi, *Vital Statistics on American Politics*, 75.

58. Seymour Martin Lipset and William Schneider, *The Confidence Gap* (New York: Free Press, 1983).

59. Thomas Patterson, *The Mass Media Election* (New York: Praeger, 1980); Michael Baruch Grossman and Martha Joynt Kumar, *Portraying the President: The White House and the News Media* (Baltimore: Johns Hopkins University Press, 1981); and Gregory Bovitz, James Druckman, and Arthur Lupia, "When Can a Liberal Reporter Lead Public Opinion?" (unpublished paper, University of California at San Diego, 1999).

CHAPTER 5

The Interpretive Media

In 1987, Gary Hart, a prominent senator from Colorado and presidential candidate considered the likely Democratic nominee, was accused of adultery. It was not exactly news that leading politicians engaged in affairs. A long line of modern American presidents from Franklin Roosevelt to John Kennedy were thought to have done so. However, in regard to Hart, it wasn't as much the sex that bothered reporters as the risk-taking character represented by the affairs. Journalists believed that this was not an aberration — that Hart repeatedly and routinely had one-night stands with many different women. This, many reporters felt, represented a different type of personality problem. In their eyes, serial affairs arose from a character flaw that revealed a deeply disturbing trait on the part of the candidate. Adulterous sex became a symbol of personal risk-taking that appeared to have important ramifications for presidential performance. The question was, if a man took dangerous risks in his personal life, would he do the same as president?

Two decades before, during the height of the objective media, reporters had downplayed adultery on the part of President Kennedy. It was an era when the mostly male journalism profession was deferential and uncritical of public officials, especially on issues related to sex. At the time, many in the press knew he had affairs with different women, but had not reported them because such behavior was thought to be "private" as opposed to "public."

Years after he had been assassinated, however, it was alleged that Kennedy's penchant for personal risk-taking had carried over into public life. Scathing exposés by a series of authors from Nancy Clinch to Nigel Hamilton, to Seymour Hersh (of Vietnam fame) claimed Kennedy had taken unacceptable risks in terms of the country's Cuban policy, dealings with organized crime, and the nascent war in Vietnam.[1] For example, Kennedy's dalliances with women associated with organized crime figures opened up the president to potential blackmail from the mob. In a related vein, secret plans to assassinate Cuban premier Fidel Castro and the actual murder of South Vietnam leader Ngo Dinh Diem pushed U.S. policy into clandestine activities that would backfire on the administration. Some even speculated that the president's ties to organized crime interests and attempts to kill Castro led to Kennedy's own 1963 assassination.

Throughout these cases of personal risk-taking, according to critics, lay a personality profile that had dangerous consequences for presidential decision making. Based on this newly emerging interpretation, personal conduct was not irrelevant to public performance or news coverage. If anything, failure to report on carousing, excessive drinking, drug taking, lying, or gambling would be irresponsible on the part of the press. It would doom America to flawed leaders who would make bad decisions.

The painful aftermath of Kennedy's assassination, anti-Vietnam war protests, and the Watergate scandal had forced everyone in society, including journalists, to reassess previous perspectives. The late-1960s and 1970s were a time of upheaval in politics, education, and journalism. Old-style news coverage based on respect for authority and deference to those in charge gave way to more aggressive reporting that probed the background of those in authority. More women and minorities were hired as journalists in order to diversify the workforce beyond white males. New questions arose: Was professional detachment and the norm of balance blinding reporters to genuine social struggles that were taking place around the world? Shouldn't journalists have a point of view that helped educate readers to what really was going on in society?

When Hart dared the press to follow him around to check his personal probity, journalists took it as a challenge to expose Hart's character flaw and inform the American public. It was consistent with the new perspective of many journalists that private behavior was relevant if it signified a major character defect; the key was the interpretation and context in which the behavior took place and what it revealed about personal character.

Slowly, the media began to evolve into another stage that was quite different from earlier epochs. Analysis and interpretation became more common. Reporters sought to bring their personal knowledge of and familiarity with public figures to the attention of readers and viewers in order to inform them about political activities. Delving deeply into personal background and character became accepted as a legitimate part of news stories. Rather than taking political statements at face value, journalists undertook analysis pieces to put particular events in a broader context.

New actors from talk radio and tabloid newspapers to news analysts and pundits gained power as professional journalists lost the monopoly they had won during the course of the twentieth century over news presentation. Having triumphed over political parties, government officials, and independent editors, reporters now lost control over the dissemination of news. Ironically, as new "journalists" appeared, the public became increasingly unable to distinguish professional journalists from their poorly trained and unprofessional colleagues. At political events, flamboyant *Rolling Stone* commentator Hunter Thompson stood side-by-side with journalistic icons such as David Broder of the *Washington Post*. The excesses of the journalistic fringe became intertwined in the public's views of the professional core, to the detriment of the latter.

The resulting alterations in news presentation fundamentally changed the nature of reporting and how citizens view the press. With an interpretive style that was more subjective and inserted journalists' own impressions and values into stories, reporters started to lose the public trust and respect they had held throughout much of the twentieth century. Rather than seeing the press as making a fair-minded effort to be balanced in its coverage, more and more people began to feel that the press was biased, unfair, and opinionated, and that reporters were probing far too intrusively into the backgrounds of candidates and public officials. Eventually, these evaluations would generate a major backlash against the press, and weaken the ability of mainstream reporters to influence the public.

THE SHIFT TO INTERPRETATION

As happened at the beginning of the objective media era, journalists in the 1980s were influenced by broader changes in a variety of academic fields. Perhaps the most important development came in the backlash against objectivity. Early in the twentieth century, journalists and academics developed great faith in facts, truth, and objective reality. The idea was that certain truths were knowable and that by ridding oneself of partisan biases and subjective filters, factual knowledge could be gained.

Such a perspective was consistent with the general notion of scientific progress that was prevalent at the time. Advances in the scientific fields of physics, biology, and chemistry spilled over into thinking about human civilization itself. If atoms could be split and genes mapped, then perhaps objective truths about human behavior could be learned as well. Die-hard adherents of this worldview believed it was inevitable that social behavior be understood as well as natural phenomena were.

Behind the popularity of this way of thinking was the growing acceptance of science in a world that previously had relied heavily on superstition and intuition. Educated people around the turn of the twentieth century believed in the power of scientific progress and the virtue of objectivity in the pursuit of knowledge. If individuals were open to rigorous scientific methods, facts could be discovered and society advanced. It was a perspective that dominated much of the twentieth century.

In the post–World War II era, though, many academic fields began to move away from objectivity as a major means of developing knowledge. Truth and knowledge were seen as relative to context, personal position, cultural values, and societal norms. New theories of social relations suggested that beneath the surface of "reality" lies a complex set of thoughts, values, and social norms that are not easily understood. Writers such as F. A. Hayek argued that facts did not exist by themselves, but were a function of how we think. "What is relevant in the study of society is not whether these laws of nature are true in any objective sense, but solely whether they are believed and acted upon by the people," he wrote.[2] Cultural biases, for example, permeate institutional arrangements and societal norms, and affect how people think about one another. Reality is embedded in larger social structures and ways of thinking.

Fields from philosophy to psychology to film and comparative literature began to look at knowledge in different ways. Eventually, whole new approaches based on "postmodern" thinking were put forward. In this context, it was not enough to try to determine what was factually correct; instead, it was critical to understand how cultural values made us feel about what we were seeing. Facts in and of themselves could not be interpreted out of context. Instead, it was important to understand events in light of deeply rooted structures.[3] Even if we were not consciously aware of these structures, they served as powerful determinants of how we perceive social and cultural life.

Slowly, these intellectual ideas began to have a discernible impact on how journalists practiced their craft. It was not sufficient merely to discover so-called objective facts. Rather, events must be placed in context and understood within the framework of broader structures. News developments needed to be probed and interpreted so that readers and viewers could understand what really was taking place. Superficial observations were no longer sufficient; coverage became more interpretive and more contextually based.

THE RISE OF NEWS ANALYSIS

One way in which journalists became more interpretive was through news analysis pieces placing stories in broader context or "instant analysis" following major speeches. Often paired on the front page with more traditional news stories or on television networks immediately after a major event, such analyses injected reporters directly into the story. Rather than reporting the facts of particular events, reporters now explained what lay behind the surface that was relevant to the news story.

The justification for this change was that reporters had unique insights into political coverage. Because they traveled with politicians and talked with them behind-the-scenes, journalists felt they were able to interpret actions in a way that informed voters about the larger political significance. Indeed, many in the press felt it was vital for reporters to exercise this unique form of oversight; putting political events in context would help educate the public about the broader currents of public life.

A review of newspaper front pages and television broadcasts reveals that outside of commentary and editorial pages, such analyses were rare until the mid-1980s. Indeed, industry professionalization at the turn of the twentieth century dictated against this very type of news presentation. It reeked too much of the heavy-handed editorializing of Pulitzer and Hearst, who loved to put their opinions directly on the front page. The whole thrust of the objective media throughout the twentieth century had been to remove this type of partisanship and personal interpretation from news coverage. It was part of the general effort to drive bias and subjectivity out of reporters' writings.

Yet, now journalists feared the glorification of objectivity. If professionalism meant removing events from their larger political context and helping viewers understand what lay beneath the surface, objective journalists were robbing citizens of valuable information. A host of academic disciplines from semiotics to literary theory to social constructivism were arguing that context was everything. Taking events out of context was as misleading as lying about facts and figures. Indeed, it was irresponsible journalism.

For example, when they wrote about the lives of extraordinary people, contemporary biographers placed these individuals in the social, economic, and political context of their times. The "great man" theory of history was giving way to an interpretation of people who at particular moments in time were able to seize the initiative and channel broader historical forces toward the accomplishment of great feats. Individuals such as Henry Ford, Charles Lindbergh, and Susan Anthony were products of their times. It was impossible to understand their individual accomplishments without reference to the world in which they lived.

In the same manner, political events had to be placed in context in order for them to be understandable. It was not enough to report the "who, what, and where" surrounding an event. Rather, reporters must delve into human motivation, explore why particular actions were undertaken, and provide some sense of the larger significance of these activities. Only in that way could someone really understand what was happening politically. In other words, the trend toward news interpretation and analysis was a response to concerns about the illusory nature of objectivity.

AD WATCHES

About the same time that news analysis pieces started appearing on the front pages, ad watches became prominent features of election coverage. These so-called "truth boxes" were designed to evaluate candidate claims in political advertisements. Rather than accepting candidate statements at face value, it was the job of reporters to critique and interpret advertising claims.

Popularized by reporters such as Brooks Jackson of CNN, Elizabeth Kolbert of the *New York Times*, and Howard Kurtz of the *Washington Post*, among others, these ad watches were designed to help voters understand the truthfulness of ads and to point out falsehoods that were present as well. Reporters would review the text of ads and note when the truth was shaded or the candidate's claim was false. On television, problematic ads would get superimposed with splashy graphics proclaiming that the commercial was "FALSE," "MISLEADING," or "UNFAIR."

Much like news analysis pieces, ad watches were designed to contextualize the claims of political candidates. Recognizing that election campaigns sometimes bring out the worst in candidates because of the need to appeal to voters in closely contested races, exaggerated claims bias an election in favor of candidates who shade the truth. It therefore is important that journalists provide an oversight function for voters.

First appearing on a wide scale in 1988, these ad watches soon became very prominent. In 1992, for example, the *New York Times* ran forty-four separate ad watches, while the *Washington Post* ran forty-five. In 1996, due to industry concern over effectiveness and a one-sided presidential campaign, the number of ad watches dropped to twenty-three for the *New York Times* and fourteen by the *Washington Post*.[4]

Ad watches were accompanied by companion pieces on television called "reality checks." Here, broadcast reporters would show an ad and critique its claims. A common approach would be to broadcast the commercial and have a reporter verbally tell viewers what was dishonest or deceptive about the spot.

However, owing to the visual nature of television, early reality checks were flawed in one important manner. Sometimes, according to focus group participants, viewers would remember the ad shown in the feature, not the critique. Or to put it differently, news coverage would amplify rather than correct misleading information contained in the spot.

This limitation led to a refinement of the technique. Rather than showing the candidate ad full-screen with audio commentary by a reporter, only portions of the spot were shown and those segments were placed in a small box at an angle on the screen so that viewers understood that the feature's goal was to contextualize the commercial. The reporter then would evaluate the ad and use visual technology to highlight segments that were particularly misleading. The idea was that broadcasters needed a visual approach in the critique that was as powerful as the ad itself. Using large type superimposed on the screen over the ad or graphics that clearly would convey to casual viewers what the problem of the ad was, these new ad watches overcame some of the limitations of early versions.

Even with these refinements, some analysts still did not like either the newspaper or television features.[5] The problem, according to these observers, was that reporters interjected themselves into the story and made subjective judgments as to what was correct or false. Critics worried that on the one hand, these evaluations would not be very effective at educating voters and on the other hand, would become a dangerous source of journalistic bias at a crucial time in the election process. From the standpoint of these observers, ad watches represented a step backward into subjectivity, not a leap forward into a more informative type of citizen education.

POLITICAL PUNDITRY

Along with news analysis and ad watches, the 1980s saw the rise of a new and controversial type of media interpretation known as political punditry. Pundits, as they were called, were individuals (often academics) whose subject area expertise allowed them to place events in a broad political and historical context. One of the things that frustrated many reporters was the superficiality of press coverage. If limited to covering objective reality, it sometimes was difficult to help voters understand developments as they were taking place.

Reporters themselves were not the ideal people to contextualize events because readers and viewers feared that journalists were being biased in their subjective judgments. Outside academics, on the other hand, had considerable credibility as political analysts. Since they were uninvolved with the political process and not seeking office, they had exactly the type of source credibility that made them successful interpreters.

Initially, the media used their own staff to analyze political events, such as CBS relying on commentator Eric Severeid or later using *Newsweek* columnist Joe Klein. Sometimes, prominent liberals and conservatives were paired in order to balance the discussion. As prominent individuals, pundits helped attract viewers interested in hearing what they had to say about contemporary political events. By dint of their high-profile platforms, these media commentators carved out an important niche for themselves in the political process.

But eventually outside academics and analysts were brought in to help interpret particular events. Scholars such as Norman Ornstein and William Schneider (later hired full-time by CNN) of the American Enterprise Institute, Thomas Mann of the Brookings Institution, Larry Sabato of the University of Virginia, and Kathleen Hall Jamieson of the University of Pennsylvania became prominent political pundits who appeared frequently on television, and in newspapers and magazines.

Pundits also appeared in other areas. Retired generals were hired to analyze military events, such as during the Persian Gulf war. In fact, after that war, NBC retained General Norman Schwarzkopf for news feature stories. Business leaders and academic experts assessed the stock market and business trends. Lawyers such as Roy Black, Gerry Spence, and Greta Van Susteren appeared frequently on television as legal commentators.

Consistent with the logic of news analysis and ad watches, these pundits helped interpret particular events. They assessed what really was going on, filled the public in on the story behind the story, and noted the broader significance of events. Proficient in the history of societal events and at ease in front of a television camera, pundits gained wide exposure as analysts.

Between 1979–80 and 1987–88, according to one study, the number of pundit appearances on the three national television networks nearly tripled from 88 to 260. This represented nearly one pundit appearance per broadcast day. The most frequent "news shaper" identified by this research project was William Schneider, with fifty-eight individual appearances in 1987 and 1988.[6]

Other studies pointed to Ornstein as the leading political pundit of the 1980s and 1990s. In looking at Ornstein's press citations from 1983 to 1995, there was a rapid acceleration in his number of newspaper and magazine citations during this period. For example, Ornstein's pundit analysis jumped from around 30 instances in 1983, to 155 in 1986, to 200 in 1989, to 300 in 1992, and 380 in 1994. The trendline in punditry clearly was increasing at a geometric rate.

However, in 1995 and 1996, there was a reshuffling of the pundit throne. At that point, Mann finally bested Ornstein as king of the pundits. From May 1, 1995, to April 30, 1996, Mann was cited 277 times in newspapers and magazines, compared to 251 for Ornstein and 191 for Schneider. Press coverage publicized this seismic shift in pundit wars, but it could not mask the overarching reality. Pundits had become a major vehicle for news analysis and interpretation throughout the industry.[7]

The reason for this increasing reliance by the news industry on pundits was clear. Studies documented that news analysts have high source credibility. When compared to other political figures, such as presidents, politicians, interest group representatives, and bureaucrats, commentators had the highest credibility and were one of the most powerful determinants of how the public responded to important news events. It was therefore not surprising that the industry shifted its coverage in favor of news analysis.[8]

By the mid-1990s, nearly every leading newspaper and television station prominently featured pundits whose job it was to interpret political events. From local stations and area newspapers to the *New York Times* and *Washington Post*, these purveyors of context and interpretation carved out an important role for themselves in media coverage.

THE WILLIAM KENNEDY SMITH TRIAL

A powerful illustration of the role of analysis and interpretation in the media came in Spring 1991, with the indictment of a member of a prominent American political family. Over Easter weekend at the Kennedy family compound in Palm Beach, Florida, William Kennedy Smith, the son of Stephen and Jean Smith and nephew to Senator Ted Kennedy, was accused of assaulting a woman he had picked up at a bar. After he met her, Smith had brought the woman back to the family's beachfront compound. She claimed he raped her at the house, while he countered that they had engaged in consensual sex.

The case exemplified the state of interpretive media because it immediately became part of the emerging national debate over "date rape," when casual encounters cross the line into sexual assault. From the standpoint of the media, the case had much greater significance than the actual lawsuit involving two people. Rather, it was an opportunity for the media to put events in the context of a much bigger social problem.

With a number of female reporters having entered the journalism profession in the 1970s and 1980s, there was attention to issues that previously had been ignored by male reporters. Date rape during this time was becoming recognized as an important social issue. Previously, the public's image of rape was a stranger grabbing a woman in a dark alley and sexually assaulting her. Yet according to the Federal Bureau of Investigation's 1991 Uniform Crime Report, only 23 percent of rapes involved strangers. In fact, rape was more likely to come at the hands of acquaintances (40 percent of all rapes), boyfriends or husbands (29 percent), or another relative (8 percent). This new portrait of sexual assault made plain that large numbers of rapes occurred at the hands of social acquaintances who refused to accept that "no means no."

In date rape, casual dates turned tragic and the legal system had to sort out competing claims of "he said, she said." In stranger assault, DNA testing could establish whether sexual penetration had taken place. But in the new world of date rape, when the man admitted sex but argued it was consensual in nature, physical evidence could not definitively establish whether the sex had been forced.

As soon as word became public that a Kennedy had been accused of assault, a media feeding frenzy erupted. Dozens of press outlets descended on Palm Beach. The networks were there, CNN used its twenty-four-hour news channel, and local stations and newspapers from all around the country sent reporters to cover the case. When the news flow was slow, pundits and analysts filled the air with commentary about the larger meaning of the trial and what the case represented politically and legally for the United States.

Emblematic of this new type of media coverage, the story was not restricted to "who, what, and where," but attempted to place the case in context. Media analysis discussed the Smith indictment within the framework of the broader social problem of date rape. What Smith was accused of was symptomatic of a harsh reality. It would not be so much Smith on trial, but the more general issue of date rape, and the complicated feelings that it generated from all sides, that would be the focal point of media coverage.

For months, the publicity percolated. The tabloid press and new television outlets such as *Hard Copy* and *Inside Edition* regaled readers and viewers with each titillating detail. Senator Ted Kennedy had been present. Indeed, he had been the one who, after a painful evening of reminiscing with his sister Jean about the death of her husband Stephen, had dragged William Smith and his own son Patrick Kennedy, a congressman from Rhode Island, out of bed to go to the bar.

With the involvement of the extended Kennedy family, coverage shifted to the history of Kennedy tragedies. The media pondered questions such as: Were the Kennedys guilty of bad judgment? Was there a bad seed in the family that explained recurring incidents? Did this weekend demonstrate that male Kennedys were boozers and womanizers who had exceeded the boundaries of good behavior?

The media event metamorphosed way beyond the details of the trial. It no longer specifically involved the individual behavior of William Kennedy Smith. Rather, the entire Kennedy family was on trial. Between date rape, the prominent Kennedy family, and a legal system grappling with a controversial social issue, the story had all the ingredients of a Shakespearean drama.

From the media's standpoint, the trial was a boon and its ratings were terrific. Indeed, the case set the tone for a string of sensational trials, all covered extensively by the press: There was Michael Jackson, who was accused of molesting young boys; the Menendez brothers, who were indicted for murdering their rich parents for money; and as discussed in Chapter 6, there was the saga of O. J. Simpson, who was tried for murdering his former wife Nicole Brown Simpson and Ron Goldman.

In the end, Smith was acquitted of sexual assault in a jury trial. Media analysts debated whether the acquittal represented the case of a "fatal attraction" — a woman trying to frame a prominent man — or evidence of a rich and powerful political family "getting off" through expensive attorneys and private detectives. The evidence from the court of public opinion was not as easy to decipher. The case aroused a wide range of reactions from people. While some felt the Kennedys had bought justice, others believed the accuser needed help and that the media were out of control in their news coverage.

Although the publicity over date rape and the Kennedys soon quieted down, the episode came to personify the new era of media coverage that was emerging. The objective media era was ending, and a new approach to news gathering and reporting was taking its place. Professional journalists were losing their monopoly over news coverage to pundits, the tabloids, and talk radio, none of whom necessarily shared the same values and training as the mainstream media. Context and interpretation were in, and objectivity was out. Personal character and background now were defined as newsworthy topics.

The interpretive media era opened the media up to controversial debates over the proper nature of reporting and whether journalists were going "too far" in their coverage. Citizens liked the broader perspective reflected by the interpretive media and the effort to contextualize political and cultural events. But they worried about its subjective and oftentimes personalistic tone. Sex was titillating, yet the public was not clear it should be splashed all over the front page and on television screens. Character reporting was a double-edged sword from the standpoint of the American public, especially as the media messenger itself became more controversial.

THE HILL-THOMAS HEARINGS

In the fall of 1991, another controversial example of interpretive reporting took place. Clarence Thomas, a conservative black Republican, had been nominated by President George Bush for Thurgood Marshall's seat on the U.S. Supreme Court. It was a bold Republican effort to appease those who wanted to maintain minority representation on the court, but move the seat in a much more conservative direction.

A native of Pin Point, Georgia, Thomas was from a poor family. After receiving a law degree from Yale University, Thomas had risen through the federal bureaucracy. Eventually, he was appointed to head the Equal Employment Opportunity Commission (EEOC), where he sought to weaken affirmative action. A sign of his empathy with the GOP, Thomas was once described by a Bush staffer as one of the "friendly minorities."[9]

Following Marshall's retirement, Bush decided to put forward Thomas's name for the Court seat. The president's own popularity was at an all-time high following the successful Persian Gulf war. Right-wing conservatives were still upset that in his previous court appointment, Bush had nominated David Souter, a man who was seen as too moderate. Bush chose to appease the right wing with Thomas's appointment. And the conservative, black man with poor Southern roots seemed an unbeatable political combination.

Liberals were uneasy about Thomas. He was unrelentingly conservative and had little experience as a judge. In public hearings, they tried to trap him into expressing views on controversial subjects, such as abortion, that would create problems for him with the general public. But Thomas did not bite. Coached by top Republican consultants, he demurred on these questions, saying he had no predetermined opinion on abortion and would decide controversial cases based on their individual merits.

Back in Norman, Oklahoma, where she taught at the University of Oklahoma Law School, Anita Hill was watching the case with unusual interest. Before accepting the teaching position, Hill had worked with Thomas at the EEOC. His nomination concerned her. Not only was she worried as an African American about Thomas's conservative philosophy, she had a vivid recollection of a more personal nature. While at the EEOC, Thomas allegedly had raised topics of a sexual nature with her. Boasting of his own physical prowess and penchant for pleasing women through oral sex, Thomas repeatedly sought dates with Hill and pressed her to have sex with him.

At this time, sexual harassment was a rarely discussed and largely unenforced crime, one that was generally ignored by the press. Less than five thousand women filed harassment complaints with the EEOC in 1991. Although many women around the country knew first-hand how often male bosses made advances to female subordinates, it was an issue that women were expected to deal with privately. Indeed, more than 90 percent of victims never filed formal complaints (a number that included Hill).[10] Though Thomas was persistent, Hill repelled his advances and looked the other way. She never reported him to authorities or filed a formal complaint. Hill didn't want to cause trouble because Thomas was in a position to help or hinder her professional advancement.

But elevating a known harasser to the Supreme Court was another matter. For weeks, she agonized over what to do. Eventually, she was contacted by a Senate Judiciary Committee staffer, who had heard rumors of the personal problems between them. The staffer encouraged Hill to go public with her story. Hill talked privately with the committee staff, and hoped that would end the matter. But before she knew it, her accusations were leaked to the press on October 6, 1991. Amidst blazing headlines and tabloid stories, she was asked to testify before the Senate Judiciary Committee on national television.[11]

It turned out to be one of the most riveting testimonies in congressional history. In excruciating detail, Hill described how Thomas had harassed her. He persistently made comments about her physical appearance. He would turn to her and say, "You look good, and you are going to be dating me, too."[12] According to her, he loved pornographic movies, especially "materials depicting individuals with large penises or breasts involved in various sex acts."[13] She cited an episode that typified the problem. Hill described "an occasion in which Thomas was drinking a Coke in his office. He got up from the table at which we were working, went over to his desk to get the Coke, looked at the can, and said, 'Who has put pubic hair on my Coke?'"[14]

By the end of her testimony, Thomas's nomination looked on the verge of collapse. On national television and before a Senate committee, a credible witness had accused a Supreme Court nominee of bizarre personal behavior and sexual harassment. Although the graphic nature of her testimony made almost everyone uncomfortable, Hill's calm and professional demeanor evoked considerable sympathy.

At this point, Thomas decided to respond publicly to the charges. Millions of Americans had heard her testimony and awaited his reply. How would he handle the explosive charges about his personal character and on-the-job behavior? Would he be able to refute the explosive charges that she had made?

On the day of his testimony, Thomas opened with a stirring defense of his character. "I cannot imagine anything that I said or did to Anita Hill that could have been mistaken for sexual harassment," he said.[15] Then, moving directly to the issue of race, always an undercurrent in the nomination proceedings, Thomas blasted the hearing as a "high-tech lynching for uppity blacks." He, for one, would not "supply the rope for my own lynching."[16] It was a clear reference to the anti-black violence of the Reconstruction period.

Throughout the live testimony of Hill and Thomas, the media provided saturation coverage of the charges and counter-charges. Reporters probed Hill's background. Was she a person who made up things, fantasized about sex, or had a grudge against Thomas? Typical of "he said, she said" cases, her personal character became as important as his. Because there were no witnesses to their office conversations, viewers' judgments came down to who seemed more honest and believable in their respective testimonies.

Polls showed considerable public interest in the case. The viewing audience for the hearings was estimated at 27 million.[17] A Gallup national survey found that 86 percent of a national sample claimed to have watched at least an hour of the hearings.[18] In the nine days between when the accusations were made public and the Senate confirmation vote on October 15, eleven different national polls were conducted by media organizations.

Overall, a majority of Americans accepted Thomas's denial of Hill's charges, according to most of these surveys. For example, an October 14 poll revealed that 56 percent thought Thomas was telling the truth and 27 percent felt Hill was being truthful. However, there was a significant gender gap between men and women. Whereas 63 percent of men in this survey said they believed Thomas, only 49 percent of women felt that way.[19] On the crucial race dimension, 70 percent of blacks expressed support for Thomas. His powerful imagery of lynching had mobilized the African-American community behind him.[20]

One of the hallmarks of press coverage during the era of the objective media was that there was little differentiation in coverage due to the background of the reporter. It mattered little if the reporter were male or female, liberal or conservative, or Republican or Democrat. In fact, Stephen Hess argues during his study of Washington reporters, that despite their overwhelming Democratic affiliations, the fundamental professionalism of the press was enough to squeeze partisanship and personal bias out of reporting.[21]

However, with the rise of the interpretive media, subjective judgment became more closely intertwined with coverage. Interpretation and analysis varies significantly with personal background. And when reporters have a role orientation that allows for interjecting their own views into news stories, it is little surprise that coverage reflects personal views and experiences.

On the Hill-Thomas hearings, male and female reporters often sparred on the air and in newspapers over the credibility of the two witnesses. The popular expression among women reporters in regard to their male colleagues was "they just don't get it." By this, female journalists felt that men simply didn't understand how these things could happen to a woman and why Hill would keep quiet and maintain a continuing relationship with the boss despite the harassment.

Studies of press coverage during the hearings revealed sharp differences in how reporters covered the event. One study of 390 news stories printed between October 11 and 14, 1991, found that reporters focused on four dramatic themes: melodrama, eroticization, personal conflict between the two individuals, and privatization of the act into the personal sphere.[22] The author concluded that "[t]hrough these dramatic press accounts, Thomas and Hill became colorless cardboad cutouts, standing in for the epic struggle between men and women everywhere." Continuing, this observer argued that "gender . . . played a major role . . . in media constructions of the hearing."[23]

Another study of the ABC, CBS, and NBC stories from October 6 to 14, 1991, dealing with the hearings revealed that "male news sources favored Clarence Thomas, while female sources supported Professor Hill." For example, Thomas earned positive ratings from six of every seven males quoted in the stories, as opposed to one of three from female sources. In contrast, 75 percent of female sources praised Hill, compared to 44 percent of male sources.[24]

In the end, the Senate voted to confirm Thomas by a vote of fifty-two to forty-eight. It was the closest vote ever held for someone who was actually confirmed to the Court. Others had attracted more negative votes than Thomas, but had not been confirmed. Despite the tortuous testimony and saturation media coverage on a controversial sexual assault charge, Thomas had received a lifetime appointment to the most important judicial body in the country.

IMPACT ON MEDIA CREDIBILITY

If media credibility reached its high point during the objective media stage, then the interpretive phase saw the start of a slow but steady decline in public trust and confidence in the media. In 1992, on the heels of the Thomas-Hill hearings, there were signs of a growing backlash against the mainstream media.

The public was deeply ambivalent about character-based reporting. On the one hand, people were titillated by cases involving date rape and sexual harassment. Public interest both in the Kennedy Smith and Hill-Thomas cases was quite high. Television, in particular, reported higher ratings during these episodes than at other times.

However, ordinary citizens were not entirely comfortable with having such deeply personal behaviors splashed onto the news. America always had been conservative on social and moral questions, and these new issues provoked a range of strong feelings. While large percentages of Americans reported that they thought the press should report on spouse abuse (71 percent), income tax evasion (65 percent), and exaggerations of military or academic records (61 percent), few believed journalists should cover personal matters such as whether a female candidate had an abortion (17 percent), the use of antidepressants (20 percent), marijuana use (23 percent), or a past affair (23 percent).[25]

Reporters were shifting from symbiotic ties to public officials, which had been common during the objective media, to adversarial relationships. This was especially apparent on the White House beat, where a series of presidents encountered distinctly unfavorable news coverage. As presidents sought to shield themselves from aggressive media oversight, they limited access to journalists, which had the effect of further alienating reporters.

The result of all these changes in the way in which journalists did their jobs came in the form of a fleeing audience. In the 1980s, the television networks lost fifteen to twenty points of audience share (see Appendix Figure A.3). And newspaper circulation levels, which peaked in 1947, continued to fall (see Appendix Figure A.1).

Beyond the size of the viewing audience, public ratings of media job performance started to drop. Whereas earlier two-thirds of the American public gave the media positive ratings for how they handled their jobs, in 1992 about half were positive.[26] Nixon and Agnew's attacks on the press had not resonated much with the public in the 1970s, but two decades later citizens were beginning to have major doubts about news gatherers. The drop in media ratings was the clearest sign of public concern about how reporters were defining the news and presenting information.

The public was becoming more worried about whether press coverage was inherently biased. When asked in the 1992 campaign whether they thought campaign coverage was biased against particular individuals, 43 percent felt it had been and 49 percent said it was not. Public faith in a fair and balanced national press corps was starting to erode. These perceptions were confirmed by later content analysis that showed of the three candidates, Bush earned the most negative press. Seventy-one percent of comments about him in network newscasts were negative, compared to 48 percent negative for Clinton and 55 percent negative for Perot. Public fears about press bias, therefore, were not unwarranted.[27]

The high source credibility of journalists, which had long been crucial to the special power exercised by the American press, was weakening. With the erosion of media credibility came a decline of their power as well. If reporters were more subjective in contextualizing information, it was easy for the public to conclude they were biased, unfair, and not to be trusted. The ability of the objective media to sway public opinion and mold people's reactions, demonstrated so clearly in earlier cases, was weakening. The image of the omnipotent media, key to their political power, was not as strong. Reporters no longer were seen in the same glowing light they had been before this era and no longer held the monopoly on news gathering.

In seeing evidence of more subjective coverage on the part of journalists, some observers concluded that reporters were returning to their roots in the partisan press. For example, Senator Jesse Helms, the well-known conservative from North Carolina, attacked the press throughout the 1980s with complaints about its liberalism. Noting academic research, which had documented that reporters were far more likely to be liberal and Democratic than conservative and Republican, he castigated the media for their so-called "liberal bias."[28]

Meanwhile, liberals for their part were equally dissatisfied with media coverage. During the Reagan administration, a number of liberal critics felt Reagan was getting a free ride from reporters.[29] The president often made wild claims, which upon fact-checking proved erroneous. But nothing the press reported seemed to dent his popularity. Many reporters appeared to give up on stories which noted still another Reagan misstatement. It was no longer considered newsworthy and, in any event, had no impact on the views of the public at large.

However, coming from opposite sides of the political spectrum, each of these critiques missed the point about the limitations of the interpretive media. There were biases in media coverage during this period, but they were not systematic distortions dictated by ideology or partisanship. Rather, the biases were episodic in nature and directed toward the sensational. Sometimes, media coverage moved in a liberal direction and was supportive of liberal spokespeople, such as Ralph Nader and Jesse Jackson. At other times, the coverage was conservative in nature and critical of liberals such as Ted Kennedy.

This inconsistency meant that over time media coverage could be favorable to Reagan in the 1980s (before Iran-Contra) and turn negative toward him when the Iran-Contra details emerged. The press could be generous to Bush in his 1988 election, but devastating toward him in 1992. Reporters could be sympathetic to Clinton in the 1992 election, but extremely negative later in his presidency.

The erratic nature of press interpretations confused the public and took a toll on media credibility. With reporters seemingly liberal one day, conservative the next, and cynical nearly every day, public confidence in the press eroded.[30] It was easy for citizens to conclude that the media had become less reliable in its judgments about public officials and that reporters no longer should be trusted to convey the news impartially. The glory days were coming to an end.

NOTES

1. Nancy Clinch, *The Kennedy Neurosis* (New York: Grosset & Dunlap, 1973); Nigel Hamilton, *JFK: Reckless Youth* (New York: Random House, 1992); and Seymour Hersh, *The Dark Side of Camelot* (New York: Little, Brown, 1997).

2. F. A. Hayek, *The Counter-Revolution of Science: Studies on the Abuse of Reason* (Glencoe, Ill.: Free Press, 1952), 30.

3. Claude Levi-Strauss, *Structural Anthropology* (Garden City, N.Y.: Anchor Books, 1963).

4. Darrell M. West, *Air Wars: Television Advertising in Election Campaigns, 1952–1996* (Washington, D.C.: Congressional Quarterly Press, 1997), 97–101.

5. Stephen Ansolabehere and Shanto Iyengar, *Going Negative: How Political Advertisements Shrinks and Polarize the Electorate* (New York: Free Press, 1995), 139–140.

6. Lawrence Soley, *The News Shapers: The Sources Who Explain the News* (New York: Praeger, 1992), 31. Also see Dan Nimmo and James Combs, *The Political Pundits* (New York: Praeger, 1992); and Eric Alterman, *Sound and Fury: The Washington Punditocracy and the Collapse of American Politics* (New York: HarperCollins, 1992).

7. Paul Starobin, "Deposing the King," *National Journal*, May 18, 1996, 1124. For a critical view of pundits, see Franklin Foer, "Quotemeisters," *New Republic*, November 4, 1996, 21–22.

8. Benjamin Page, Robert Shapiro, and Glenn Dempsey, "What Moves Public Opinion?" *American Political Science Review* 81 (1987): 23–43.

9. Jane Mayer and Jill Abramson, *Strange Justice: The Selling of Clarence Thomas* (Boston: Houghton Mifflin, 1994), 18.

10. "Keep Up the Campaign to Stop Sexual Harassment," *USA Today*, October 15, 1992, 10A.

11. Anita Hill, *Speaking Truth to Power* (New York: Doubleday, 1997); and Timothy Phelps, *Capitol Games: Clarence Thomas, Anita Hill, and the Story of a Supreme Court Nomination* (New York: Hyperion, 1992).

12. Mayer and Abramson, *Strange Justice: The Selling of Clarence Thomas*, 132.

13. Ibid., 108.

14. Ibid., 291.

15. Ibid., 290.

16. Ibid., 299.

17. Dianne Rucinski, "Rush to Judgment? Fast Reaction Polls in the Anita Hill-Clarence Thomas Controversy," *Public Opinion Quarterly* 57, (Winter 1993): 575.

18. Larry Hugick, "On Night before Vote, Support for Thomas Remains Strong," *Gallup Poll News Service* 56 (October 15, 1991): 1–4.

19. Dianne Rucinski, "Rush to Judgment? Fast Reaction Polls in the Anita Hill-Clarence Thomas Controversy," 590. Also see Virginia Sapiro and Joe Soss, "Spectacular Politics, Dramatic Interpretations: Multiple Meanings in the Thomas/Hill Hearings," *Political Communication* 16 (July–September 1999): 285–314.

20. Dan Thomas, Craig McCoy, and Allan McBride, "Deconstructing the Political Spectacle: Sex, Race, and Subjectivity in Public Response to the Clarence Thomas/Anita Hill 'Sexual Harassment' Hearings," *American Journal of Political Science* 37 (August 1993): 699–720.

21. Stephen Hess, *The Washington Reporters* (Washington, D.C.: Brookings Institution, 1981).

22. Lisbeth Lipari, "As the Word Turns: Drama, Rhetoric, and Press Coverage of the Hill-Thomas Hearings," *Political Communication* 11 (July–September 1994): 299–308.

23. Ibid., 301.

24. John Carmody, "Sources Exhibit Sex Bias," *Providence Journal*, October 21, 1991, C5.

25. Pew Research Center for the People & the Press, "Too Much Money, Too Much Media Say Voters," September 16, 1999 press release, <www.people-press.org>.

26. West, *Air Wars: Television Advertising in Election Campaigns*, 184–185. Also see Alex Kuczynski, "Another Glum Portrait of American Journalists," *New York Times*, December 28, 1998, C6.

27. Marion Just, Ann Crigler, Dean Alger, Timothy Cook, Montague Kern, and Darrell M. West, *Cross Talk: Citizens, Candidates, and the Media in a Presidential Campaign* (Chicago: University of Chicago Press, 1996), chapter 5.

28. Dan Nimmo and James Combs, *The Political Pundits* (New York: Praeger, 1992), 146–151.

29. Mark Hertsgaard, *On Bended Knee: The Press and the Reagan Presidency* (New York: Farrar, Strauss, & Giroux, 1988).

30. James Fallows, *Breaking the News: How the Media Undermine American Democracy* (New York: Pantheon, 1996); and Thomas Patterson, *Out of Order* (New York: Vintage Books, 1993).

CHAPTER 6

The Fragmented Media

Early in 1992, Bill Clinton was the clear front-runner for the Democratic presidential nomination. With President Bush's approval ratings hovering near 90 percent in 1991 following the successful Persian Gulf war, a number of leading Democrats had opted not to seek the presidency out of fear that the election was a lost cause for their party. In the weak field of Democrats, Clinton appeared to be the strongest candidate.

But out of the blue in early February, two weeks before the crucial New Hampshire primary, the *Star* broke a story that threatened to derail his entire candidacy. A former television anchorwoman from Arkansas named Gennifer Flowers claimed Clinton had engaged in a decade-long affair with her. Armed with an audiotape of a recent phone conversation she had with Clinton, Flowers played the tape during a press conference broadcast live on CNN. Soon, every news outlet in the country was covering the story in lurid detail.[1]

From a media standpoint, the most interesting thing about the incident was that the allegation had already been known to many professional reporters at major news outlets. Clinton opponents had been shopping the story around to mainstream reporters without much success. On the heels of the Gary Hart-Donna Rice saga, news scribes were not eager to force another Democratic candidate out of the presidential race through a sex scandal. The public had grown weary of character assaults by the press. After a year devoted to William Kennedy Smith's and Clarence Thomas's respective sexual travails in 1991, the mainstream press decided that Flowers's charges were not newsworthy.

In the objective era, when a handful of elite outlets dictated the agenda and pack journalism was in vogue, that decision would have killed the story. With top journalists concluding that the story was not newsworthy, it would not have received much attention. The charges simply would have faded away. Similar to FDR's mistress and JFK's adultery, a conspiracy of elite media silence would have protected Clinton from public disclosure.

But in the 1990s, a deregulated era was dawning that would usher in a stage known as the *fragmented media*. In this period, we would see a new role for the local news, the rise of new networks, a tabloid press, talk radio, satellite technologies, the World Wide Web, and the passage of the 1996 Telecommunications Act designed to spur competition and deregulate the industry. In their totality, these changes would dramatically increase the number and diversity of news channels available to the public, weaken pack journalism, and further undermine the public's sense of press professionalism that had governed their impressions for nearly a century.

The days when ABC, CBS, NBC, the *New York Times*, and the *Washington Post* could dominate news gathering would give rise to tabloid journalism, cutthroat media competition, heterogeneous coverage, and an era when peripheral press outlets could break major news stories as easily as the elite press. The professional monopoly over news gathering and presentation that had been held by mainstream journalists for most of the twentieth century would decline precipitously. The rise of alternative new sources would have major consequences for the way news got covered, the relationship between the press and public officials, and the degree of confidence citizens had in the media. Press fragmentation would mark the fall of the media establishment as a powerful and unified entity. With weak public esteem and declining public respect, reporters would lose the credibility and homogeneity of coverage that had given them such great influence during the objective era.

FRAGMENTATION OF THE MEDIA MARKETPLACE

The mass media in the 1960s and 1970s were an elite club dominated by a small number of leaders and a patrician sense of responsibility for the industry. Between the two wire services, three national television networks, and a handful of top newspapers, the press operated under a "follow-the-leader" mentality. Opinion leaders at top outlets defined the news and determined what stories were worthwhile. Lacking prestige and resources, other news organizations generally followed the agenda set by the industry leaders.[2]

During this period, the industry was an institution as important as Congress and the presidency. Politicians cultivated journalists in order to generate favorable coverage.[3] Observers used to think of the press as a haphazard group of people with no organization to guide individual behavior. But this portrait missed regularities in news routines and the way in which professional norms and organizational structures guided the industry.[4]

However, beginning in the 1980s and accelerating in the 1990s, the elite media club would be destroyed through the rise of new technologies and new competitors in the news marketplace. Along with talk radio, music videos, local news, and the World Wide Web, new networks and a profusion of cable channels would expand the viewing and listening options available to citizens — changes that dramatically altered the industry.

The clubby world of ABC, CBS, NBC, and PBS gave way to nearly one hundred television channels. All-news channels such as CNN, MSNBC, and CNBC offered political coverage around-the-clock, whether or not there was breaking news. New television networks appeared, such as Fox, UPN, and WB. Television alternatives like talk radio skyrocketed in popularity. Local news stations moved aggressively into coverage of national political events. And beginning in 1991, the World Wide Web began to transform the way people accessed information.

The impact of this virtual revolution in media structure was staggering. Not only were there dozens of channels on many cable systems, the country now had a profusion of national television networks (seven at last count), talk radio shows (four thousand), newsletters (more than 1 million), radio stations (over twelve thousand), Internet sites (more than 1.5 million), and magazines of various sorts (over twenty thousand). Paraphrasing the old Chinese saying, many more than "one thousand flowers" bloomed in the Information Age.

The result of all these changes on the elite television networks was dramatic. Appendix Figure A.3 demonstrates that the mass audience among all households with televisions for the "big three" networks fell from just under 60 percent in 1976 to 27 percent in 1999. If one looks only at the television-watching segment of the audience, the network share fell from 90 percent in 1976 to 43 percent in 1999. As CBS News President Andrew Hayward noted in 2000, "you have an increasing fragmentation of the media into niches."[5]

Viewers are now as likely to watch news shows on cable television as on broadcast networks. For example, 60 percent of Americans claim they regularly watch one of the cable news networks (such as CNN, CNBC, MSNBC, Fox News Channel, ESPN, or the Weather Channel), while 57 percent say they watch the major broadcast networks' news shows regularly. As a sign of public fickleness, half of those who watch the news do so with a remote control in their hand.[6] Freed of the need to walk across the room to change channels, Americans switch stations at the first sign of boredom. It is a technological flexibility that breeds little loyalty in viewer behavior.

It therefore is unsurprising that with a dramatic increase in media options, people have deserted the networks in droves. The prime-time competition no longer is ABC, CBS, and NBC, but ESPN for sports lovers, old comedy shows on Nick-at-Nite, variety options on the Comedy Channel and the Family Channel, and *Star Trek* on UPN, among many others. It is an era of cutthroat competition for viewers and ad revenues. Even a hit show like Fox's *Ally McBeal* is watched by only 10 percent of America's households with television. And a smash success like *Seinfeld* with extraordinary ratings by contemporary standards barely would have made the top 20 in the 1970s.[7]

Speaking to this decline in network viewing audience, ABC President Robert Iger said, "We used to think the possibility existed that the erosion was going to stop. We were silly. It's never going to stop. As you give consumers greater and greater choices, they are going to make more choices."[8]

If the core media was previously defined as a few prestige organizations, today the media includes a bewildering variety of outlets, such as the tabloids, *Inside Edition*, and the talk radio shows of Don Imus, Howard Stern, and Rush Limbaugh, among others. Limbaugh's radio show, which started in 1988, is carried by 650 stations across the country and reaches an estimated 15 million listeners a week. Featuring conservative rhetoric combined with entertaining satires on leading political figures, Limbaugh brought a tabloid-style mentality to talk radio. Overall, more than four thousand talk shows were broadcast on 1,200 radio stations, representing a ten-fold increase over the past two decades.[9]

It is not just radio that has exploded. Desktop publishing programs have stimulated a flurry of newsletters and magazines. According to industry sources, in 1998, there were over 1 million free newsletters published in the United States, around 18,000 subscription newsletters, and nearly 20,000 magazines.[10]

In the 1970s, the only television news magazines were CBS's *60 Minutes* and ABC's *20/20*. Competition was strong, but genteel. In following decades, all that would change as the number of news magazines increased dramatically. The latter shows were joined by *48 Hours, PrimeTime Live, Turning Point, Day One, Dateline, NOW, Saturday Night, Face to Face*, and *Eye to Eye*, along with the syndicated tabloid shows: *Hard Copy, Inside Edition*, and *A Current Affair*.

As demonstrated by the Gennifer Flowers story, peripheral outlets such as the *Star* or *Hard Copy* can break news stories as easily as the *New York Times*. Even CNN is capable of launching an independent presidential candidacy, as it did with Ross Perot in 1992 when he gained prominence and declared his election bid on the *Larry King Live* show.

In this environment, the competition became intense and upset the clubby atmosphere of the old media. As William Small, the president of NBC News, pointed out, "In the old days, there was a pecking order. If you represented the *New York Times*, doors flew open. If you were a crusader, you wanted to appear on *60 Minutes*. . . . Now, TV has eclipsed most of print with all these magazine programs. There is competition for all these interviews like never before."[11]

THE ROLE OF MURDOCH AND TURNER IN FRAGMENTING THE MEDIA

Similar to earlier eras, when prominent people such as Horace Greeley, Joseph Pulitzer, and William Randolph Hearst transformed the mass media of their day, many of the shifts in the contemporary period arose when a new set of maverick owners wanted to significantly change the industry. New technological developments such as cable television, satellite technology, computers, and desktop publishing clearly facilitated this revolution, but as happened in previous stages of American history, several key individuals had the insight to harness new technologies and reshape the news industry.

One such man was Rupert Murdoch. This maverick did not share the patrician view of William Paley of CBS, Adolph Sulzberger of the *New York Times*, or Katherine Graham of the *Washington Post*. Each of these individuals were establishment creatures who were comfortable with and indeed benefited from the media status quo. They were at the top of the industry pecking order, and as such, exercised tremendous power over what got covered and how things were reported.

Murdoch, on the other hand, was an irascible conservative bent on challenging the business and political establishment.[12] Born in Australia, he had gotten his start in the media business by inheriting a small newspaper from his father. From this base, Murdoch had taken over papers in Perth and Sydney, and created Australia's first national newspaper. When these ventures proved successful, he branched out into television stations in Australia, New Zealand, and Hong Kong. His big international break came in the late 1960s when he purchased two London tabloids, the *Sun* and *News of the World*. He used the money from these profitable ventures to acquire the *New York Post, Boston Herald, Chicago Sun-Times,* the *London Times, TV Guide,* HarperCollins book publishers, the Twentieth Century Fox studio, and Metromedia television stations. The latter formed the nucleus of his new American television network, Fox Broadcasting.[13]

Murdoch's modus operandi was to buy a marginal property, run tabloid-style coverage, squeeze the competition, and make a lot of money. A specialist in cutting costs, he loved to stretch the limit of what was considered in good taste to attract readers and viewers. His formula proved to be exceedingly successful. Cash generated from each new media property allowed Murdoch to expand further and take on establishment icons over several different continents. In this regard, he was the first true global news mogul, with holdings all around the world.

In each new city where he bought a media outlet, the technique was the same. He ran tabloid-style coverage, exposés of scandals involving prominent people, and stories to entertain the masses. Similar to the philosophies expressed by Pulitzer and Hearst at the turn of the twentieth century, Murdoch loved to appeal to ordinary people by going after political, social, and economic elites. After buying the British newspaper the *Sun,* for example, Murdoch predicted in his very first editorial that the newspaper "will never, ever sit on fences. It will never, ever be boring." To the consternation of more conventional media outlets, these were promises he kept.

In his first issue of the *Sun* on November 17, 1969, Murdoch ran a lead story headlined "Horse Dope Sensation/*Sun* Exclusive" that dealt with a horse racing scandal. Along with this story was a picture of a Swedish model (clothed in this issue— nude photos later became daily features), an interview with British Prime Minister Harold Wilson, and excerpts from Jacqueline Susann's novel, *The Love Machine.*

Murdoch's combination of "sex, fun and sensationalism" proved quite successful. Within a year, the *Sun*'s circulation had doubled. He was on his way to confounding critics who laughed at his low-brow antics and ridiculed his long-term prospects.[14]

A decade later, when Murdoch entered the U.S. market, his approach remained the same. In the summer of 1977, after Murdoch had purchased the *New York Post,* a New York assailant with a .44 pistol started a killing spree. Eventually, the serial murderer would acquire the name Son of Sam. The editor of the *Post,* Steve Dunleavy, whom Murdoch had brought in from the *Star,* made an open appeal to the murderer to personally surrender to him. When the police arrested a suspect named David Berkowitz, the *Post* headline screamed "CAUGHT!" in bright red. Sales of that edition more than doubled to over 1 million.[15] The next day, reporters from the *Post* were arrested for breaking into Berkowitz's apartment; nevertheless, the *Post* ran a story headlined "INSIDE THE KILLER'S LAIR," complete with pictures of the killer's abode.

From the beginning, when Murdoch had promised to take clear stances, his papers played an active role politically. The publisher's politics were conservative and antiestablishment. Nothing pleased him more than taking on liberal icons and subjecting them to scorn and ridicule. From the *Boston Herald*, Murdoch eviscerated liberal Democratic Senator Ted Kennedy. Calling him "fat boy," Murdoch's columnists made sure that each Kennedy scandal from Chappaquiddick to the Palm Beach rape trial got repeated and hostile coverage.

In a move that echoed the tabloid press of the late nineteenth century, Murdoch ran editorials endorsing particular candidates on the front page of his newspapers. Following each high-profile endorsement, news coverage would be blatantly biased in favor of the chosen candidate. Opposition figures would be slammed and the paper's favorite extolled. An example of this came in the 1978 New York City mayoral race between Ed Koch (Murdoch's preferred candidate) and Mario Cuomo. An analysis of front-page stories found articles that "were far more favorable to Koch than to any other candidate. Indeed, there were quite simply no unfavorable stories about Koch."[16] With help from Murdoch's newspaper and television holdings, Koch won the primary and was then elected mayor of New York. It was not the last time Murdoch befriended a conservative politician. Murdoch used his control of major communications outlets to help friends and punish enemies.

In 1995, his publishing company negotiated a controversial book deal with new Republican Speaker Newt Gingrich. In exchange for rights to Gingrich's next book, HarperCollins promised the Speaker a book advance of several million dollars. The tycoon denied that the various regulatory issues his company had before the federal government played any role in the large advance.

Murdoch was not the only maverick who transformed the news industry in the 1980s and 1990s. Down South, a man named Ted Turner also was confronting and beating old-line media leaders. Like Murdoch, Turner was an iconoclastic individual who had gotten his start in the business by inheriting an outdoor advertising business from his father (although in Turner's case, his namesake had committed suicide). By age thirty, Turner had his first million dollars. Later he joked that the rest of his wealth came easily after that first million.[17]

But in other ways, Turner was quite different from Murdoch. Turner had a tough time growing up. Reminiscent of Hearst's problems as an undergraduate at Harvard, Turner had flunked out of Brown University. From the beginning when Turner had announced to his two roommates that he was "the world's best sailor and the world's best lover," it was a mismatch between him and the university.[18] Turner never took his classes seriously and drank far too heavily. When he was caught with a woman in his room, a flagrant violation of university rules, he was unceremoniously kicked out of school. Turner never finished his college degree, although later he was awarded an honorary degree from the university.[19]

Turner was also quite different from Murdoch in his politics. Unlike the Australian, Turner was far more liberal and interested in issues such as the environment and population growth. But both were contrarians who dreamed big and saw unparalleled opportunities in new technology for challenging old-fashioned and backward-thinking media titans.[20]

In Turner's case, it was cable television, satellites, and an Atlanta superstation that formed his power base. Superstations were local channels that through cable systems could parlay sports interests and old movies into a nationwide audience.

Turner built a rundown local station WTCG in Atlanta into a nationally prominent superstation — WTBS — with a large viewing audience. WTBS specialized in broadcasting Atlanta sports teams and old movies. It did not run any news, in recognition of Turner's belief that "no news is good news."[21]

From money earned in that venture, Turner made history, ironically enough, by founding the country's first around-the-clock news station. CNN started broadcasting in 1980. Initially, Turner wanted to have "a news program for children" on CNN in the afternoon. His business associate, Reese Schonfeld, protested saying, "We're not going to have any kid's show in the middle of the afternoon. That doesn't work." Schonfeld's plan for CNN included "world news, local news, Washington news, financial news, interviews, and as much live coverage as they could afford." Turner was puzzled by this plan. He explained, "what you're saying, that sounds a lot like . . . it's gonna be like regular news." Schonfeld defended his proposal saying the network had to go with something that was a known commodity.[22] Eventually, the two forged an agreement on CNN as an all-news channel, and started the process of challenging the networks.

Critics mocked the new network as "Chicken Noodle Network," an amateurish operation from its personnel to broadcast conduct. Its employees were "young, inexperienced, [and] underpaid." The only star on its roster was Daniel Schorr as CNN's Washington correspondent.[23] The first broadcast was not a very good omen; within the first hour, CNN lost its national feed and went off the air temporarily.

Over the years, though, CNN became a huge success, surpassing the wildest expectations even of its founders. Turner had started a radical new concept in television broadcasting, the twenty-four-hour, all-news format. With its live coverage of breaking news and interviews with leading news makers, CNN would compete effectively with the networks and spawn many imitators. As the world shifted into the new era of a fragmented media marketplace, CNN would be present at every major event, giving the American public a front-row seat at a dizzying array of national and international occasions. From major trials and congressional testimony to live press conferences and the start of bombing in Baghdad during the Persian Gulf war, CNN would reshape the news and turn the industry on its head.

THE RISE OF THE WORLD WIDE WEB

While Murdoch and Turner were transforming the newspaper and television industries, respectively, a quiet but equally significant development was unfolding in the computer arena. From humble origins in the hands of industry leaders such as Thomas Watson, Steve Jobs, and Bill Gates, among others, the computer industry grew to revolutionize information access and news dissemination. Taking advantage in the early 1980s of tiny microprocessing chips that were capable of storing enormous amounts of information, the personal computer allowed people to access and process information more quickly than ever before in the history of man. Envisioned initially as independent devices that would allow individuals to work more efficiently by themselves, computer companies soon introduced networking capabilities that allowed people on computers to talk with one another along shared information networks.

Although the industry grew through the efforts of a number of talented people, no one personified the new computer era better than Gates, the founder of Microsoft Corporation. Dedicated to creating new software that would make computers easy to use, Gates did as much as anyone to expand the popularity of the industry. He specialized in creating software for common applications such as word processing and spreadsheets.

Similar to Hearst and Turner, Gates was a college dropout. After enrolling at Harvard University in 1973, Gates spent most of his time there skipping classes and playing cards. His poker games often lasted much of the night, and Gates sometimes found himself finishing a hand at 8:00 A.M.

With his childhood friend Paul Allen, Gates was one of the first to see the business potential of the computer revolution. Together, the two started Microsoft in 1975 in Albuquerque, New Mexico. That location was chosen because it was the home of MITS, the first company to market inexpensive computers to the public. While others developed the software, Gates focused on "sales, finance, and marketing." By 1979, the company had moved outside of Seattle and was negotiating million-dollar contracts. In 1980, Microsoft got its big breakthrough when it started working with IBM on the development of a personal computer.[24]

Over the course of two decades, Gates became the world's richest man. By 1999, his personal fortune was estimated at $90 billion. Microsoft had become the dominant software distributor in the world. With its lucrative revenues, the corporation acquired other companies and moved into new computer niches.

In 1991, a technological innovation took place that was as important to the news industry as the telegraph had been in 1844 and coast-to-coast television broadcasting in 1946. A computer scientist named Tim Berners-Lee at the European Particle Physics Laboratory in Switzerland developed a technology that would allow people anywhere on the planet to access text, pictures, and sounds simultaneously via computer linkages.[25]

Gates initially was slow to see the value of this innovation but soon grasped its significance. Combined with software developed by Netscape and Gates's Microsoft Company, Berners-Lee's dramatic breakthrough made possible the World Wide Web, a computer network that has revolutionized the transmission of information. According to industry experts, at the beginning of 1998, there were "30 million computers in the system, 70 million users (worldwide), 1.5 million WWW servers, serving 350 million web pages of information." Overall, the size of the Web in 1999 was put at 3 trillion bytes of information.[26]

Each day, thousands of new Web sites are created, which turns nearly every American into a virtual broadcaster. A Pew Center national survey in 1996 found that 21 million Americans (12 percent of the voting-age population) said they obtained political information on-line sometime during the year. Seven million said they had used the World Wide Web specifically to obtain information about the presidential campaign that year.[27]

By 1998, the number getting news at least once a week from the World Wide Web had risen to 20 percent of all adults, or 36 million people in all. Not surprisingly, young people were the ones most likely to go on-line; 30 percent of those age eighteen to twenty-nine said they went on-line at least once a week; 24 percent of those age thirty to forty-nine did; 13 percent of those fifty to sixty-four did; and just 4 percent of those age sixty-five or older went on-line. And in 1999, 41 percent of Americans indicated they were regular Internet users.[28]

Not only was the Web affecting where people got the news, it was transforming how the industry presented coverage. A survey of 192 newspaper editors and 170 magazine editors revealed that between 1995 and 1998 the number of outlets with a Web site doubled from 25 to 58 percent. More and more journalists also were posting original accounts on their newspaper's Web site, as opposed to merely duplicating stories already available through their print editions.[29]

One of the things consumers found attractive about the World Wide Web was its unscriptedness and spontaneity. Whereas mainstream media outlets had editors and fact checkers, the World Wide Web operated on the stream of consciousness principle. Whatever someone thought could appear almost instantly on computer screens around the world. When a TWA plane went down off the Long Island coast, the World Wide Web spun a fanciful series of conspiracy theories involving the government shooting down innocent civilians.

While the total news audience on the World Wide Web grew by leaps and bounds, no individual site dominated the market. According to industry estimates, in March 1998, the country's two largest news Web sites, CNN (founded by Turner) and MSNBC (a union between Microsoft and NBC) attracted 4.2 million and 3.3 million viewers, respectively, on any given day. Even with several million viewers, the CNN audience share of all Web site viewers was only 8 percent and MSNBC was just 6 percent.[30] At this point in time, the World Wide Web looked more like the FM radio dial with lots of small stations, each of which had tiny listening audiences.

But the Web's overall ability to attract audiences was starting to grow into a major force. Companies included their Web site address in the newspaper and television advertisements that they ran. The sale of products through the Web skyrocketed. More and more people were going on-line to obtain news, and buy goods and services. In 1999, the top Web sites in terms of monthly visitors were America Online (47 million), Microsoft (32 million), Lycos (31.9 million), Yahoo (31.3 million), Go Network (23.8 million), Geocities (21.3 million), Excite (18.9 million), Time Warner Online (13.3 million), Blue Mountain Arts (11.1 million), and Amazon (10.7 million).[31]

The gentlemanly rivalry that characterized the oligopolistic media in the 1960s was devolving into one of chaos and cutthroat competition. Structural fragmentation and deinstitutionalization became the new world media order. With thousands of new information options from the tabloids to talk radio to the World Wide Web, old organizational routines were being transformed. The former media were not becoming extinct but were facing a horde of new information competitors.[32] A variety of new outlets were starting to gain a foothold in people's information networks.

At the same time that new outlets were arising, waves of layoffs and buyouts robbed mainstream media outlets of their most professional and experienced staff. A press that had been governed by professional norms for news gathering and layers of bureaucratic checkpoints to ensure accuracy was losing its structure. Unfettered market competition was turning every media outlet and every reporter into direct competitors.

New York Times reporter James Bennet concluded, "With the Internet and cable television fracturing the audience, wonderfully diversifying the coverage, and providing more opportunities for direct access to audiences, I wonder how the mainstream media can stay relevant."[33] Only the fittest would survive this brutal competition — the electronic equivalent of armed combat.

THE DECLINE OF HOMOGENEOUS NEWS COVERAGE

The most dramatic consequence of the changing media structure was in the type of coverage that was becoming prevalent. In the previous era, news coverage had a homogeneous tone. A variety of reporters covered the same event and reached similar conclusions about what was newsworthy. Professionalism created a continuity in the coverage that was designed to ensure that people heard the same story, regardless of the journalist or news outlet involved.

Indeed, critics complained vociferously about the "herd" mentality on the part of journalists. Writing about the 1970s, Timothy Crouse condemned "pack journalism" in his best-selling book *The Boys on the Bus*.[34] It didn't matter whether the reporter worked at the *New York Times* or the *San Francisco Examiner*, professional norms and a shared sense of what was newsworthy kept reporters producing the same type of coverage. And if that didn't work, the follow-the-leader mentality in the media ensured that elite outlets would guide the coverage of other news organizations.

This homogeneity was most apparent in coverage of presidential campaigns. For example, research on the 1972 campaign found virtually identical coverage across all three of the television networks. Looking at the amount of coverage as well as degree of issue discussion and ideological slant, researchers found no real differences in how the campaign was reported.[35] A follow-up study of the 1976 campaign reached basically the same conclusion. After examining media coverage at several different outlets, Thomas Patterson concluded that "although the press is not monolithic in how events are reported, it is in which events are covered."[36]

Homogeneity also tended to be the norm early in the 1990s. One in-depth study of the 1992 presidential campaign found differences in amounts of news coverage between network news, local television news, and newspapers, but a similar tone.[37] For example, whereas the *Boston Globe* and *Los Angeles Times* devoted more than 22,600 inches of space to the 1992 presidential campaign, the *Fargo (N.D.) Forum* devoted 5,600 inches and the *Winston-Salem Journal* in North Carolina devoted 9,364 inches. The same was true for local television. WCVB in Boston devoted nearly twenty hours of coverage to the campaign, while KABC in Los Angeles broadcast nine hours, WDAY in Fargo devoted three and one half hours, and WXII in Winston-Salem ran three hours of coverage.[38] Despite differences in volume of coverage, the tone was generally similar across media outlets. Analyzing the coverage on a five-point scale running from 1 (negative) to 5 (positive), the tone of reporting fell within the range of 2.43 to 3.17 for Bush and 2.96 to 3.55 for Clinton. ABC was slightly "warmer" in its coverage of Clinton and CNN was a little more positive toward Bush.[39]

However, another research project on the 1992 campaign revealed "there [were] significant and consistent differences among media on important democratic dimensions: voice, substance, and cynicism."[40] Comparing coverage by newspapers, national television, and local television, researchers found that newspapers presented the most detailed policy information, local television news the least. In terms of degree of cynicism in the coverage, newspapers and national television were much more cynical than local television news.

Of course, these studies were early in the development of the fragmented media and focused only on the mainstream media. Had the research been broadened to include Limbaugh's talk radio show, *Hard Copy*, newsletters, magazines, and the World Wide Web, greater disparities in reporting surely would have been uncovered. The range of organizations covering the campaign was infinitely diverse and the coverage reflected this heterogeneity of news outlets. Limbaugh, for example, was unremittingly hostile toward Clinton, much more so than the mainstream press. The voluble talk radio host lampooned both Bill and Hillary Clinton for their liberalism and, in Limbaugh's eyes, their shaky personal values. In contrast, hosts such as Geraldo Rivera were more positive toward Clinton.

By the 1996 presidential election, differences among individual reporters began to become apparent. According to media monitoring studies by the nonpartisan Center for Media and Public Affairs, which reviewed 573 ABC, CBS, and NBC evening news stories about the primary campaign, CBS's Eric Engberg was the most critical reporter with 87 percent of his stories negative in tone, followed by NBC's Bob Faw (87 percent critical), NBC's Lisa Myers (86 percent negative), and CBS's Bob Schieffer (79 percent critical). At the other end of the spectrum, NBC's Tom Brokaw was most positive, with 54 percent of his stories being favorable in tone, and CBS's Joe Klein had a 50 percent positive rate.[41]

In the late 1990s, even more discernible differences across major media outlets opened up on a range of foreign and domestic policy events. According to the Center for Media and Public Affairs, "the three network news broadcasts gave the public three different views on the Kosovo story."[42] Drawing on 208 ABC, CBS, and NBC newscasts during the first eight days of U.S.-NATO action, researchers found that ABC was "highly critical," CBS was "heavily supportive," and NBC "fell in the middle of the spectrum." Whereas 63 percent of statements in ABC news stories were critical of American policy, only 33 percent of CBS comments were unfavorable, as was 46 percent of NBC's coverage. There also were substantial differences in how network television handled the President Clinton-Monica Lewinsky scandal. In reporting on this story, "NBC was the most likely to quote unnamed sources (72 percent of all stories), followed by CBS (52 percent) and ABC (49 percent)."[43]

The same development has been documented by political scientist John Woolley in a study of media coverage of child abuse. Examining the reporting of several national newspapers — the *New York Times, Los Angeles Times, Christian Science Monitor, Washington Post,* and the *Wall Street Journal* — from 1982 to 1997, he found only a modest correlation between news outlets in the number of stories reported each year by the five papers.[44]

As new media actors enter the business of reporting and commenting on the news, the norm of similar coverage based on widely shared professional norms is breaking down. Due to the increasing competitiveness of news outlets and the concomitant loss of industry professionalism among reporters, journalists are not producing the same coverage. Consequently, what people watch and read often determines the type of story they hear.

The rise of interpretative journalism in the 1980s allows differences in coverage based on the background of reporters into news analysis pieces. Freed of the constraints of objectivity, viewers and readers have started to see significant differences in coverage based on the sex and race of the reporter. Combined with the unscripted setting of talk radio and the World Wide Web, the fragmented media has raised reporter subjectivity and bias to new heights in the news industry. In the process, it has created serious problems for journalists on several high-profile news cases.

THE O. J. SIMPSON TRIALS

One vivid example of this startling transformation within the mass media came in June 1994, when O. J. Simpson, a former National Football League running back and movie actor, was accused of murdering his wife Nicole Brown Simpson and Ron Goldman. With accusations of a gruesome and bloody murder by a prominent celebrity, hordes of reporters descended on Los Angeles for the nationally broadcast trial.[45]

In the competitive new world of journalism, there was virtually no self-restraint in the coverage. CNN and MSNBC discussed every aspect of the case for months on end. Days were filled with legal jockeying, discussions of new developments, and analysis by an army of legal pundits. National polls assessed in minute detail whether ordinary Americans believed Simpson was guilty of murder.

The case had all the elements of a good soap opera: sex, race, big money, and celebrity status. Simpson was black and a former football star, while his slain wife was a beautiful, blond, white woman. After going through a succession of lawyers, Simpson settled on famed lawyer Johnnie Cochran for his defense. The story riveted America in a way no other case had, with the exception of the Lindbergh baby kidnapping case in the 1930s. CNN saw its biggest jump in the ratings since the 1991 Persian Gulf war, when it had exclusive pictures from Baghdad, Iraq, at the start of the American bombing. CNN's audience increased five-fold, to nearly 2.3 million homes a day.[46]

Despite protests about the sheer volume of coverage by television and newspapers, individuals followed the proceedings with great interest. Everyone had an opinion about the Simpson case based on information gleaned from news coverage. The trial became an important part of mass culture. Male and female reporters argued on the air about domestic violence as did white and minority journalists about the fairness of the legal system.

Like the William Kennedy Smith and Clarence Thomas scandals before, the Simpson episode became emblematic of a larger social issue, in this case domestic violence. In 1994, according to figures from the U.S. Department of Justice's Bureau of Justice Statistics, over 900,000 women were assaulted by current or past spouses or boyfriends.[47] When it was revealed that Simpson had beaten his wife to the point where she was bruised, and the two had argued constantly during and after their marriage, pundits debated what this case told us about American society. The event took on a larger meaning than just a murder trial, becoming a symbol of how the United States was changing near the turn of the millennium. New issues from date rape to sexual harassment to domestic violence were roiling the country, and complicating the ability of journalists to report the news professionally. On such emotionally charged subjects, it was difficult for even the most well-intentioned reporters to provide coverage that was fair and balanced.

As in the era of the interpretive media, significant differences in news coverage of the trial emerged depending on which outlet was read or watched. Many mainstream newspapers ran a series of negative stories about Simpson prominently featuring assertions of guilt. For example, within a week of the killing, the *New York Times* printed an overview article citing Los Angeles County District Attorney Gil Garcetti's claim that evidence in the case would prove "beyond the necessary burden of proof" that Simpson was guilty of murder. The only unresolved question, according to the DA, was whether the prosecution should "ask for the death penalty." The story noted that Simpson was deeply depressed and in jail under a suicide watch.[48] The implication was that he was guilty.

In sharp contrast, NBC's *Today Show* was much more cautious about proclaiming Simpson's guilt. Hosted by Bryant Gumbel, a black man who had risen through the news industry from an early career covering NFL football, this show, with its average daily viewing audience of 14 million, appeared more favorable to Simpson. Cochran was a periodic guest on the show to explain Simpson's perspective on the case. Gumbel came across as personally sympathetic to the accused football star, whom he had known from his days covering the NFL.

Black-operated newspapers, with their national readership of 13 million, meanwhile turned Simpson into a victim of the white-controlled legal establishment. For example, the *Philadelphia Tribune* ran a front-page story with the banner headline, "The Assassination of O. J. Simpson: High-Tech Bounty Hunters Score — Again." Barnett Wright, the managing editor at that paper, characterized white press coverage as "the breaking of another black man." An editorial in the *Chicago Crusader* explained the angry white press reaction to Simpson by pointing out that "white men have a deep abiding fear that black men will take their women from them."[49]

A researcher who studied black press coverage of the Simpson case found that most of it focused on five questions: "Is Simpson worthy of blacks' sympathy given his lack of involvement with anything black since his football-playing days ended? Was he singled out for prosecution because of his preference for white women and because the murder victims were white? [Had] the mainstream media replaced the old lynch mob in destroying a black man perceived to have violated racial taboos? [Was] he the latest victim of a racist society's conspiracy to destroy black men? Can any black man, even one as wealthy as O. J., get a fair trial?"[50]

These variations in news coverage across different outlets were matched by a huge race gap in national public opinion polls. Whites overwhelmingly felt that Simpson was guilty of murdering his wife; blacks did not. Not only did blacks doubt Simpson's guilt, they were much more likely to hold negative views of the legal justice system and to believe that courts and the police were biased against African Americans in general.

When the trial got underway, news reporting of the proceedings was so plentiful that it wiped away coverage of almost all other news. Candidates in California political races, for example, complained that they simply were unable to attract any news coverage to their campaigns because of the inordinate news space devoted to the Simpson case.

Unlike former news eras, people could judge for themselves what they thought. The trial was broadcast live on national television and viewers could hear the testimony firsthand. There was much less journalistic filtering than usual. Everyone could be their own media editor in this case and evaluate the primary material for themselves. To a certain extent, this freed ordinary people from the confines of media-initiated interpretations. Viewers could be more autonomous in their news judgments.

The trial created new celebrities. Individuals such as Kato Kaelin, who lived in a guest house on the Simpson estate, became nationally prominent; Kaelin even went on to host his own cable television show. People from limousine drivers to police detectives to next-door neighbors became household names for many Americans. New pundits specializing in legal maneuvering appeared on the networks. Every move and countermove by the various attorneys involved in the case was dissected on television: Should the prosecution have allowed Simpson to attempt to put a blood-soaked leather glove on over latex protectors? When Simpson was unable to squeeze the glove on, his attorney proclaimed, "if the glove doesn't fit, you must acquit."

The biggest bombshell, though, came at the end of the criminal trial, when the nearly all-black jury unanimously acquitted Simpson of the murder charge. For weeks, people debated the verdict. Had the jury let Simpson off because of the race card played by Cochran, when he asked jurors to send the police a message against minority harassment? Was justice served? Media pundits vociferously argued the merits of the trial as well as the broader meaning of the case for American society.

This was not the conclusion of the episode, however. Following the criminal trial, Nicole Brown Simpson's family along with Fred Goldman, Ron Goldman's father, sued O. J. Simpson in civil court for the wrongful deaths of the two people. The trial before a nearly all-white jury in suburban Los Angeles produced a guilty verdict and $33.5 million judgment against Simpson. Again, pundits debated the role of race in the jury deliberations. Was the only difference in the two verdicts the racial composition of the jury?

By the conclusion of the two trials, nearly everyone was dissatisfied with media performance in the case. Although viewers had encouraged news coverage by watching shows and buying papers devoted to the trial, many Americans claimed the press had spent far too much time on the trial and had contributed to the circus atmosphere surrounding the case. It was a clear example of the declining respect Americans had for their news media.

THE CLINTON SEX SCANDALS

As the Simpson case unfolded, a new spectacle appeared that would shock the nation even further. In 1994, an unknown woman from Arkansas, Paula Jones, sued the president of the United States for improper sexual conduct. According to the allegation, while governor of Arkansas, Clinton had state police troopers summon the woman to his hotel room in Little Rock after he had noticed her at an event. While in the room, according to court papers, he had exposed himself and asked Jones to perform oral sex on him.[51] If this claim had appeared in isolation, it probably would have been accorded little significance. However, as just the latest charge in a long series of criticisms about Clinton's character, this lawsuit took on greater meaning for reporters and the general public.

For years, opponents had charged Clinton with violating numerous moral and ethical precepts. Both he and his wife Hillary were accused of shady practices on an Arkansas real estate deal known as Whitewater. Gennifer Flowers claimed she had an affair with Clinton in Arkansas, a charge he was forced in 1998 to concede was true. Opponents in the 1992 campaign claimed he had dodged the draft while a college student. He was accused of illegally firing employees in the White House travel office. Critics argued Clinton systematically had abused campaign finance rules through the Democratic party in 1996.

The lengthy series of allegations took their toll on public perceptions of Clinton. Even in 1992, when he had been elected president, voters doubted his personal veracity and character. However, fear over a weakening economy coupled with Clinton's remarkable ability to communicate on a personal level with the electorate allowed him to sidestep the character issue and gain the presidency. But the doubts raised at that time and by later charges arising from Jones and the campaign finance scandals cast doubt on Clinton's personal integrity.

As the Jones lawsuit wound its way through the courts, her lawyers searched for evidence to support their theory that there was a pattern of exploitative behavior underlying the chief executive's career. His proposition of Jones, according to them, was not an isolated incident, but rather reflected his long-term propensity for extramarital affairs. They sought additional evidence to support their lawsuit.

In January 1998, they appeared to hit paydirt. A White House intern named Monica Lewinsky was accused of having a sexual affair with the president in a small hallway just off the Oval Office. According to the charge, the young woman had become friends with the president and gained access that was quite unusual for a college intern. Although every intern was given strict warnings to keep their distance from the president, Lewinsky managed to have several personal meetings with Clinton, exchanged packages and presents with him, and engaged in several dozen phone conversations with him.[52]

Shortly thereafter, another bombshell exploded. Kathleen Willey, a White House volunteer and Democratic campaign contributor, claimed on *60 Minutes* that Clinton had made unwelcome sexual advances to her when she had visited him to seek career assistance. According to her public statements, the president had kissed her and groped her breast. The entire experience had personally shocked her and not been something she had encouraged in any way, she said.[53]

The series of sex scandals provided the World Wide Web with an opportunity to become a new star in the information galaxy. An unlikely 31-year-old man named Matt Drudge had created a Web site called the Drudge Report from his small apartment in Hollywood, California. The report, authored solely by Drudge, dedicated itself to disseminating information on the growing scandal, as well as a wide variety of political and media commentary. Drudge would scoop the mainstream press with the Lewinsky accusations and eventually would be hired by ABC Radio to host his own talk show.[54]

On January 17, 1998, at the beginning of the Lewinsky affair, Drudge's Web site was averaging 80,000 visits a day. By January 21, after the intern disclosures, this number jumped to just under 220,000 hits a day. On January 24, the number of visits had skyrocketed to around 400,000 and stayed around that level until January 30, at which point it dropped back to around 220,000 daily visits for the first week in February. In the month of February, Drudge claimed a total of 6.7 million visitors to his Web site.[55]

The Clinton sex scandals opened up new possibilities for unconventional political commentators, such as comedians. According to one study, "Jay Leno, David Letterman, Conan O'Brien and Bill Maher had told 729 jokes about Mr. Clinton's sex life in the first five months of 1998 (compared with about 250 in all twelve months of 1997)." A writer for Maher's popular ABC show, *Politically Incorrect*, noted that "you now can use old dirty jokes and call them jokes about the President. It's lowered our standards."[56] For example, one Leno joke about Clinton following the president's 1999 State of the Union address was, "Clinton's speech lasted 77 minutes, which is the longest the president has ever gone without sex."[57] This blurring of the lines between politics and entertainment was noteworthy because 1996 campaign polls had shown that a quarter of Americans relied on popular culture figures like Leno and Letterman for information on the presidential campaign.

With intense competition from a variety of World Wide Web news disseminators, talk radio, tabloid outlets, and political humorists, even the mainstream press was drawn into questionable coverage on this story. According to a study by the Committee of Concerned Journalists, one-third of all stories in major news outlets (the television networks, CNN, *Time*, *Newsweek*, broadcast news magazines, and leading newspapers) during the first five days of the Lewinsky scandal were based on anonymous sources. Forty percent of these early reports consisted of "analysis, opinion and speculation, not factual reporting." In addition, 20 percent of news stories were "unverified by the news outlet reporting it and instead it was taken from some other news outlet."[58]

These kinds of tabloid tactics were becoming popular in the era of the fragmented media. The mainstream press would wait for tabloid outlets to report some juicy tidbit, which the elite press deemed too sensational for its own outlets. Such hesitancy allowed prestige journalists to claim they were above tabloid reporting. But as soon as a few outlets picked up the story, then the mainstream press would rush to cover the rumor, even if there was no independent verification. Before long, the feeding frenzy would be in full force, and there would be saturation coverage by every news organization in the country.[59] It was a trend that had been repeated many times before from William Kennedy Smith to Clarence Thomas to O. J. Simpson. Clinton merely was the latest recipient of this trend. Core outlets laundered rumors from the tabloids into coverage by mainstream reporters.

Many times, in fact, prestige outlets declined to print rumors only later to pick them up after an out-of-the-way news organization had broken the story. One example occurred in 1998 when major companies such as the *Boston Globe, Los Angeles Times, National Journal,* and the *Hill,* among others, were approached with information that House Judiciary Committee chair Henry Hyde had engaged in an adulterous affair with a Chicago beautician named Cherie Snodgrass between 1965 and 1969. Each declined to print the story. Eventually, though, an on-line publication called *Salon* broadcast the story under the title: "'This Hypocrite Broke Up My Family': The Secret Affair of Henry Hyde, the Man Who Will Judge President Clinton." According to the story, which Hyde confirmed to be true, he indeed had a relationship with a married woman three decades before. However, the committee chair argued that "the statute of limitations has long since passed on my youthful indiscretions." Immediately after *Salon* published the story, every news outlet in the country broadcast the story, often with commentary noting how unsavory it was for *Salon* to have judged the story newsworthy.[60]

Shortly thereafter, *Hustler* magazine publisher Larry Flynt stunned the political world by disclosing that incoming House Speaker Robert Livingston had engaged in at least four adulterous affairs. Flynt had uncovered this information after taking out full-page ads offering $1 million to anyone with details on sexual impropriety by members of Congress. Within hours, every major news organization in the country was publicizing the story, which prompted Livingston to resign his Speakership and retire from politics.

A *Washington Post* survey found that 40 percent of Americans "approved of what Larry Flynt is doing in revealing extramarital affairs by Republicans." Fifty-seven percent disapproved. On the delicate issue of whether the mainstream press should report the names of legislators purported to have had affairs, 46 percent said yes and 52 percent indicated no. Interestingly, according to the poll, "women disapproved of Flynt's conduct by an almost 2-to-1 margin, 64 percent to 34 percent. But men were almost evenly split, with 48 percent approving and 50 percent disapproving." And whereas 48 percent could identify Flynt as the person "paying for information on infidelities by members of Congress," only 19 percent correctly identified Chief Justice William Rehnquist as the individual in charge of the Senate impeachment trial.[61]

By the end of the Senate impeachment trial, tabloid outlets like the *Star* and the Drudge Report were circulating a decade-old rumor from Arkansas that Clinton had an illegitimate black son by a prostitute. After finding the woman, the *Star* paid her an amount "in the low six figures" for DNA material from her and the son. Comparisons with Clinton's genetic sample on file with Special Prosecutor Ken Starr later revealed Clinton was not the father of the thirteen-year-old boy.

In a similar vein, Fox News Channel broadcast a story claiming that Clinton had raped a woman named Juanita Broaddrick in 1978 and "coerced her into denying it under oath."[62] The story was part of Fox's consistently more conservative coverage, reflecting the philosophy of its irascible owner Rupert Murdoch. NBC actually sent its correspondent Lisa Myers to Arkansas for an interview with the woman, but decided not to air the story until after the Senate trial was decided. The delay led Fox news anchor Brit Hume to wear a button on the air declaring "Free Lisa Myers." When NBC did broadcast the interview with Broaddrick on February 24, 1999, the episode attracted more than 10 million viewers, about 10 percent more than normal, and stimulated a flurry of stories from every leading newspaper in the country.[63] According to the woman, who was in town attending a conference, then-Attorney General Clinton raped her in a Little Rock hotel room. An hour later, Broaddrick's roommate returned to the room and found her crying, her lip black and blue, and her clothes torn. The two immediately drove back to their hometown. The president's lawyer, David Kendall, denied the charge and proclaimed that "any allegation that the president assaulted Mrs. Broaddrick more than twenty years ago is false." Others pointed out she had filed an affidavit in the Paula Jones lawsuit denying Clinton had made "unwelcome sexual advances."[64]

The entire scandal revealed how much the news business had changed. Americans now gathered information from all sorts of new and old media outlets. In the fall of 1998, for example, CNN officials announced that its Web site traffic had jumped 50 percent the day of the president's unprecedented four-minute television confession of an affair with Lewinsky. MSNBC observed a similar ratings boost with visits twice the weekday norm. Media organizations immediately dubbed these increases the "Monica effect."[65]

The homogeneity of news coverage was starting to decline. Every evening, CNBC had a show such as Chris Matthews's *Hardball* which regularly bashed the president. His average 1998 ratings of 552,000 households were double that of 1997, when an average of 252,000 tuned in to the show.[66] Meanwhile, an MSNBC show hosted by Geraldo Rivera staunchly defended Clinton. A study of media coverage of the Senate impeachment trial vote across newspapers, television, news magazines, and the Internet found that while much of the coverage was similar, Web sites provided the most detailed, interactive, and dynamic coverage.[67] Much as had occurred during the O. J. Simpson trial, media outlets sometimes chose sides and built their ratings accordingly. In a narrowcasting universe, that was how talk shows attracted viewers. Blandness was out, while attitude and edge were in.

In terms of the broadcast networks, the Clinton scandals far outnumbered reporting on other topics. For example, ABC, NBC, and CBS devoted nearly forty-three hours in 1998 to Lewinsky coverage, compared to a few hours each for the Iraq weapons inspection dispute, the war in Kosovo, the 1998 elections, and John Glenn's space flight.[68]

Despite the unrelentingly negative media coverage of Clinton from the Internet, network television, talk radio, cable outlets, and major newspapers, the president not only continued to do well in the polls, his public support actually went up in the spring, based primarily on the strong economy and the absence of war. According to national polls, two-thirds of Americans approved of the job he was doing as president. A significant number continued to have major doubts about his personal character, but these concerns did not interfere with their generally positive view of his job performance.[69]

Even after Clinton was forced in grand jury testimony to concede that a sexual affair had taken place with Lewinsky and the House impeached him on charges of perjury and obstruction of justice, his standing in the polls remained positive. Eventually, the Senate voted not to remove the president from office. Even more impressive, in the face of hostile press coverage of a Democratic president, Democrats gained seats in the 1998 midterm elections, confounding pundits and political experts alike.

Clinton's strong public support provided him with crucial political insulation. Seizing the moment, Clinton sought to shift blame for the negative coverage back to the media. In one speech, the president noted that he was too busy to read newspapers much anymore, but that he did find time to read all of the retractions. It was a not-so-veiled jab at the press for having reported several aspects of the sex scandals erroneously, and having to print corrections shortly thereafter.[70]

THE DEATH OF JOHN F. KENNEDY, JR.

The tragic plane crash and death of John F. Kennedy, Jr., his wife, and sister-in-law on July 16, 1999, is the most recent example of a fragmented media that careens from endless saturation coverage of the same topic to divided public views of the media industry as a whole. Struggling to deal with the larger meaning of the event, reporters resorted to common popular stereotypes of Camelot and Kennedy as "America's prince." Replete with royal metaphors and using his death as another reminder of the tragedies that have befallen the Kennedy family, the coverage illustrated the difficulties facing twenty-four-hour news organizations when there is high interest in an event, but little news to report.

When the plane piloted by Kennedy failed to land at the Martha's Vineyard airport, reporters had hours of airtime to fill without any significant information about what had happened or why the plane had vanished. As the Coast Guard launched a massive search and rescue mission, stations from CNN and MSNBC to the national television networks used all the accoutrements of contemporary reporting from pundits to polls. Friends of the family, business collaborators, trained pilots, and academic experts dissected the event, despite the dearth of information. So strapped for real news, NBC reported the results of a survey indicating that 22 percent of Americans believed there was a curse on the Kennedy family, while 72 percent did not.[71]

When the plane was found and the bodies recovered, analysts shifted to topics such as the safety of private planes, JFK Jr.'s piloting skills, the plight of the Kennedy family, and the unfairness of youthful deaths. CBS news anchor Dan Rather recited the lyrics of Camelot as tears welled up in his eyes. Barbara Walters of ABC compared Kennedy's tragic death to the car crash that killed Princess Diana. NBC Washington Bureau Chief Tim Russert noted that Kennedy was garnering exactly the type of excessive media coverage that he abhorred in life.[72]

However, befitting the heterogeneity of a fragmented media, Web site discussions of the tragedy were free-flowing and unrestrained by the niceties of professional newscasts. An America Online chat room, for example, featured no-holds-barred debates over Kennedy's role in the accident. One person with the moniker Jude861 commented that Kennedy "did nothing for anyone but him and he killed his wife and sister with his poor judgment." Fellow chatter Dolphinsue immediately objected to this characterization, saying "it was a horrible accident."[73]

For a media marketplace plagued by fragmentation and declining audience share, the tragedy was a godsend, dramatically raising the ratings of twenty-four-hour news channels across the board. For example, New England Cable News reported for the weekend of the crash that its ratings had tripled. CNN (viewed by 1.42 million households), MSNBC (viewed by 600,000 households), and Fox News (viewed by 504,000 households) reported their highest audience ever, besting previous marks set by the car crash that killed Princess Diana and the Columbine high school shootings.[74]

The time pressures and need for scoops in the new media configuration made it very difficult for various outlets to report the same versions of the news.[75] The desire for exclusives generated several inaccuracies from some stations, such as early reports that four people were aboard the downed plane (not the three who actually were) and that the bodies were recovered before this actually had happened. Each error forced retractions a short while later, leading one correspondent to complain on the air amidst all the confusion, "details of the story seem to change all the time."[76]

The saturation coverage from all outlets, regardless of the content, led media critic Tom Rosenstiel, director of the Project for Excellence in Journalism, to complain, "in twelve hours of coverage, there were only about ten minutes' worth of actual facts."[77] According to the Center for Media and Public Affairs, "the ABC, CBS and NBC evening newscasts ran 27 percent more stories on Kennedy's death (131 pieces) than they did during the first week of Diana's passing."[78]

The rise of the tabloids and the Internet necessitated one somber move on the part of the Kennedy family. In a request designed to prevent publication of grisly autopsy photographs, relatives asked that no photos be taken. According to the *Cape Cod Times*, the family's stated rationale for making this unusual request was its fear "that such photographs could land in tabloid newspapers or on the Internet."[79]

In the end, a Gallup survey found the public divided on key aspects of the reporting. On the one hand, 58 percent complained that the amount of media coverage of the Kennedy-Bissett deaths was excessive. At the same time, however, 57 percent reported a favorable view of the tone of the media's coverage, compared to 39 percent who felt unfavorably about it.[80]

BACKLASH AGAINST THE MEDIA

By the end of the twentieth century, there was widespread public unhappiness with the media. Near the conclusion of the Senate impeachment trial of President Clinton, only 35 percent indicated they approved of the job done by the media.[81] The trust and confidence Americans had expressed in the industry just two decades earlier was gone, and major concerns were being expressed about the nature of contemporary press coverage.[82] The adversarial relationship that previously had been evident between reporters and public officials now has spread to the public's views of the press.

In the 1996 presidential campaign, Republican nominee Bob Dole begged his audiences to "rise up" against the media. At his election stops, he proclaimed, "We've got to stop the liberal bias in this country. Don't read that stuff. Don't watch television. You make up your own mind. Don't let them make up your mind for you. We are not going to let the media steal this election. The country belongs to the people, not the *New York Times*."[83]

In regard to the Lewinsky scandal, a Pew Research Center national survey in March 1998 found that 55 percent rated press reporting on the scandal as only fair or poor, while 42 percent gave the media excellent or good marks. Interestingly, the group that was most negative on the news coverage was senior citizens. Among this group, only 30 percent gave the media a favorable rating.[84] Among the top complaints about the media cited in this poll were that the press "sensationalize[d] the scandal," "report[ed] information before getting all the facts," relied on "hearsay or unsubstantiated information," and showed a "continued hyping of the scandal story."

Even more disturbing was the long-term decline in views about the favorability of news organizations. From 1991 to 1998, according to the Pew Research Center polls, the media's favorability rating dropped from 91 to 76 percent. Between early and late February 1998 alone, there was a drop of seven percentage points in favorable ratings of media coverage about the Lewinsky scandal.

When asked about investigative journalism in general, national surveys found less enthusiasm for the watchdog role played by the media in the 1990s, compared to attitudes that existed in the 1980s. The days of American journalism when Woodward and Bernstein became national heroes for exposing a corrupt president, and a number of reporters won Pulitzers for their coverage of the Vietnam War, were over.

Between these decades, the public expressed increasing concern about press inaccuracy, sensationalism, and lack of fairness. In 1985, 55 percent claimed news organizations generally got the facts straight and 34 percent believed that stories often were inaccurate. By 1997, these numbers were reversed: Thirty-seven percent felt news organizations got their facts straight and 56 percent believed stories often were inaccurate.[85] This was a stunning turnabout in public attitudes toward the news industry.

People also reported mixed feelings about common investigative reporting techniques that emerged in the past decade. Just 52 percent said they approved of reporters running stories with unnamed sources, 42 percent approved of using hidden cameras and microphones, 31 percent supported reporters not identifying themselves as reporters, and 29 percent approved of paying informers for news information. Another study revealed that 67 percent of the general public believes journalists "invent stories" and 87 percent feel they "use unethical or illegal tactics."[86]

Some argued that public dissatisfaction was idiosyncratic to particular news stories. If the public was upset about the Lewinsky story, it may have been because of the undue attention to sex. Or if citizens saw the Simpson trial as being covered too extensively, it was because that case was sensationalized due to the racial aspects of the trials and the celebrity status of O. J. Simpson. Or if there were media overkill surrounding JFK Jr.'s death, it was due to the celebrated status of the Kennedy family.

However, what these interpretations miss is the way the structural fragmentation of the news media makes the kind of public dissatisfaction seen on these cases a likely part of future impressions of the news industry. With numerous outlets and cutthroat competition among them, the fragmented nature of the media marketplace means there will always be excesses in the coverage and heterogeneous reporting, that some organization is going to get the facts wrong, and that biases are going to be visible in the press coverage.[87] There simply is no way for the media to police themselves given the diversity of news outlets that exist in America.

As shown graphically in a recent Florida story, when a court justice posted on the Web pictures of murderer Allen Lee Davis shortly after he was executed in order to make the point that electrocution was barbaric, professional journalists no longer control the dissemination of news.[88] The line between trained reporters and others in a position to communicate with the general public has disappeared, making nearly anyone with access to a Web site a virtual journalist. The problem is not necessarily one of poor intentions on the part of journalists, though sometimes that is a factor. The structural competition that exists within the industry ensures that many of the things people dislike about contemporary news coverage are here to stay. If anything, they may actually get worse in the future as more outlets appear and less professional news disseminators gain importance.

To the extent that the public turns off the news or no longer trusts coverage, the media loses its influence over citizens' beliefs. The days when the media could make or break politicians are over; the high source credibility of the objective media was idiosyncratic to that era. There is little reason to expect the media to wield great influence when reporters are held in such low public esteem, when there are major differences in professionalism between the media core and periphery, and when coverage is not uniform in tone. The more subjective and opinionated reporters become, and the less able the public is to distinguish mainstream journalists from the tabloid press, the less influence news reporters hold over the political process.

NOTES

1. Randolph Ryan, "The Devious Love Tape," *Boston Globe*, February 1, 1992, 23.

2. David Halberstam, *The Powers That Be* (New York: Knopf, 1979).

3. Leon Sigal, *Reporters and Officials* (Lexington, Mass.: D. C. Heath, 1973); and Michael Grossman and Martha Kumar, *Portraying the President* (Baltimore: Johns Hopkins University Press, 1981).

4. Timothy Cook, *Governing with the News* (Chicago: University of Chicago Press, 1998).

5. Andrew Hayward, panel discussion, Brown University, Providence, R.I., February 17, 2000.

6. Lawrence Mifflin, "Poll Shows Cable News Catching Up," *New York Times*, June 8, 1998, D7. See full report on Pew survey at <http://www.people-press.org>. Also see Richard Davis and Diana Owen, *New Media in American Politics* (New York: Oxford University Press, 1998); and Bill Carter, "Shrinking Network TV Audiences Set Off Alarm and Reassessment," *New York Times*, November 22, 1998, A1.

7. Frank Rich, "What the Tube Is For," *New York Times Magazine*, September 20, 1998, 55–56. For a discussion of niche broadcasting, see Ronald Grover, "Must-See TV for Left-Handed Men under 30," *Business Week*, December 14, 1998, 104. Also see Matthew Baum and Samuel Kernell, "Has Cable Ended the Golden Age of Presidential Television?" *American Political Science Review*, 93 (March 1999): 99–114; and Richard Reeves, *What the People Know: Freedom and the Press* (Cambridge, Mass.: Harvard University Press, 1998).

8. Quoted in Bill Carter, "As Their Dominance Erodes, Networks Plan Big Changes," *New York Times*, May 11, 1998, A1. Also see Bill Carter, "TV Networks Are Scrambling to Deal with Era of New Media," *New York Times*, May 17, 1999, A17.

9. Figure cited in *Journal of Sport & Social Issues* (May 1998). Also see Wayne Munson, *All Talk: The Talkshow in Media Culture* (Philadelphia: Temple University Press, 1993).

10. The number of total newsletters comes from Newsletters Only, a national publishing company for newsletters; see their Web site, <http://NewslettersOnly.com>. The number of subscription newsletters comes from the Newsletter Publishers Association of America, which publishes the *Oxbridge Directory of Newsletters*. The number on magazines comes from the *National Directory of Magazines*, a publication of Oxbridge Communications in New York City.

11. Connie Chung, "The Business of Getting 'The Get': Nailing an Exclusive Interview in Prime Time," Discussion Paper D-28, Joan Shorenstein Center on Press, Politics, and Public Policy, April 1998, 4.

12. Thomas Kiernan, *Citizen Murdoch* (New York: Dodd, Mead, 1986).

13. William Shawcross, *Murdoch* (New York: Simon & Schuster, 1993), 15–16.

14. Ibid., 121–122.

15. Ibid., 143–133.

16. Ibid., 144–145.

17. Ted Turner, lecture at Brown University, April 26, 1995.

18. Robert Goldberg and Gerald Jay Goldberg, *Citizen Turner: The Wild Rise of an American Tycoon* (New York: Harcourt Brace, 1995), 56.

19. Ibid., 81.

20. Porter Bibb, *It Ain't as Easy as It Looks: Ted Turner's Amazing Story* (New York: Crown, 1993); Hank Whittemore, *CNN: The Inside Story* (Boston: Little, Brown, 1990); and Goldberg and Goldberg, *Citizen Turner*.

21. Goldberg and Goldberg, *Citizen Turner*, 159.

22. Ibid., 237–238.

23. Ibid., 260

24. Bill Gates, *The Road Ahead*, rev. ed. (New York: Penguin, 1996), 43–52.

25. Tim Berners-Lee, with Mark Fischetti, *Weaving the Web: The Original Design and Ultimate Destiny of the World Wide Web by its Inventor* (San Francisco: Harper San Francisco, 1999); and Elizabeth Corcoran, "On the Internet, a Worldwide Information Explosion beyond Words," *Washington Post*, June 30, 1996, A1.

26. See testimony by Vinton G. Cerf, senior vice president of Internet ArchCommerce Subcommittee, on Telecommunications, Trade, and Consumer Protection. Also see Gates, *The Road Ahead*. The 1999 figure comes from George Johnson, "Searching for the Essence of the World Wide Web," *New York Times*, April 11, 1999, Week in Review section, 1.

27. Pew Research Center for the People & the Press, "News Attracts Most Internet Users," December 16, 1996 press release, 1.

28. Mifflin, "Poll Shows Cable News Catching Up," D7. See full report on the Pew survey at <http://www.people-press.org>. The 1999 figure is reported by Jerry Ackerman, "Internet Users Grow More Diverse," *Boston Globe*, January 15, 1999, A1.

29. Matthew Rosenberg, "Web Publications Break Away from Print," *New York Times*, March 1, 1999, C13. But see Richard Davis for an interesting argument that the Internet has not transformed the dominant political power of traditional news organizations in *The Web of Politics* (New York: Oxford University Press, 1999).

30. Saul Hansell, "Eye Catching: How New Media Are Racing to Become the Mass Media," *New York Times*, May 11, 1998, D1, 4.

31. Laurie Flynn, "New Ratings Give Lycos a Reason to Celebrate," *New York Times*, April 26, 1999, C4.

32. Robert Blendon, John Young, Mollyann Brodie, Richard Morin, Drew Altman, and Mario Brossard, "Did the Media Leave the Voters Uninformed in the 1996 Election?" *Harvard Journal of Press/Politics* 3 (Spring 1998): 12.

33. James Bennet, "Siamese Twins," review of *Governing with the News* by Timothy Cook, *Washington Monthly*, January/February 1998, 58–60. Also see Christopher Harper, *And That's the Way It Will Be: News and Information in a Digital World* (New York: New York University Press, 1998).

34. Timothy Crouse, *The Boys on the Bus* (New York: Random House, 1973).

35. C. Richard Hofstetter, *Bias in the News: Network Television Coverage of the 1972 Election Campaign* (Columbus: Ohio State University Press, 1976).

36. Thomas Patterson, *The Mass Media Election* (New York: Praeger, 1980), 100.

37. Marion Just, Ann Crigler, Dean Alger, Timothy Cook, Montague Kern, and Darrell M. West, *Cross Talk: Citizens, Candidates, and the Media in a Presidential Campaign* (Chicago: University of Chicago Press, 1996), chapter 5.

38. Ibid., 93.

39. Ibid., 107–108.

40. Marion Just, Ann Crigler, and Tami Buhr, "Voice, Substance, and Cynicism in Presidential Campaign Media," *Political Communication* 6 (January-March 1999): 38.

41. Center for Media and Public Affairs, "Study Finds More Bite Than Meat in TV Election News," April 15, 1996, <www.cmpa.com>.

42. Center for Media and Public Affairs, "Networks Split on Kosovo Coverage," April 7, 1996, <www.cmpa.com>.

43. Center for Media and Public Affairs, "Scandal News Not So Bad for Bill," February 24, 1998, <www.cmpa.com>.

44. John Wooley, "Using Media-Based Data in Studies of Politics," *American Journal of Political Science* 44 (January 2000): 161.

45. Jeffrey Toobin, *The Run of His Life: The People v. O. J. Simpson* (New York: Random House, 1996).

46. Tom Lowry, "Cable Loses Juice Post-OJ," *New York Daily News*, October 5, 1995, 90.

47. Diane Craven, *Sex Differences in Violent Victimization, 1994* (Washington, D.C.: U.S. Department of Justice Bureau of Justice Statistics, September 1997), NCJ-164508.

48. Drummond Ayres Jr., "The Simpson Case: The Overview. Prosecutor Sees Simpson Case as 'Solid' One," *New York Times*, June 20, 1994, A1.

49. E. R. Shipp, "OJ and the Black Media," *Columbia Journalism Review* 33 (November/December 1994): 39–42.

50. Ibid. Also see Darnell Hunt, "(Re)affirming Race: 'Reality,' Negotiation, and the 'Trial of the Century,'" *Sociological Quarterly* 38 (Summer 1997): 399–422.

51. Tim Weiner, with Neil Lewis, "How Legal Paths of Jones and Lewinsky Joined," *New York Times*, February 9, 1998, A14.

52. "A President Deep in Trouble: The Allegation That He Had an Affair with a White House Intern — and Then Urged Her to Lie — Swept Washington and the Nation," *Minneapolis Star Tribune*, January 22, 1998, 1. Also see Francis Clines, "Testing of a President: The Accusers. Jones' Lawyers Issue Files Alleging Clinton Pattern of Harassment of Women," *New York Times*, March 14, 1998, A1.

53. John Henry, "Willey Accuses Clinton of Lying about Incident; Former White House Volunteer Claims the President Groped Her," *Houston Chronicle*, March 16, 1998, A1. Also see Michael Isikoff, *Uncovering Clinton* (New York: Crown, 1999).

54. Francis Clines, "Gossip Guru Stars in Two Roles at Courthouse," *New York Times*, March 12, 1998, A25. Also see Lawrie Mifflin, "ABC Radio Hires Internet's Drudge," *New York Times*, July 9, 1999, A13.

55. For the January Web site numbers, see Claire Shipman, "In the Belly of the Media Beast," *George Magazine*, April 1, 1998, 62. The overall February numbers come from Clines, "Gossip Guru Stars in 2 Roles at Courthouse," A25.

56. Janny Scott, "Speaking the Unspeakable (No Blushing Is Required)," *New York Times*, June 6, 1998, A20.

57. Howard Kurtz, "Americans Wait for the Punch Line on Impeachment," *Washington Post*, January 26, 1999, A1.

58. Felicity Barringer, "Study Finds More Views Than Facts," *New York Times*, February 19, 1998, A14; and Bill Kovach and Tom Rosenstiel, *Warp Speed* (New York: Century Foundation Press, 1999).

59. Kovach and Rosenstiel, *Warp Speed*; and Larry Sabato, *Feeding Frenzy: How Attack Journalism Has Transformed American Politics* (New York: Free Press, 1991).

60. William Powers, "Issues and Ideas," *National Journal*, September 19, 1998; printed from <www.cloakroom.com>.

61. Howard Kurtz, "Hooray for Larry Flynt," *Washington Post*, January 25, 1999, C1.

62. Woody West, "DNA Tests on Prostitute's Son Revive Old Story," *Washington Times*, January 11–17, 1999, National Weekly Edition, 1; and Bill Sammon and Frank Murray, "The Clinton Story That's Too Hot to Handle," *Washington Times*, February 8–14, 1999, National Weekly Edition, 1.

63. Lois Romano and Peter Baker, "Another Clinton Accuser Goes Public," *Washington Post*, February 20, 1999, A1; Howard Kurtz, "Long-Simmering Story Goes Mainstream," *Washington Post*, February 20, 1999, A9; Felicity Barringer and David Firestone, "On Tortuous Route, Sexual Assault Accusation against Clinton Resurfaces," *New York Times*, February 24, 1999, A16. The ratings numbers for *Dateline* come from Lawrie Mifflin, "All-News TV Loses Viewers after Trial," *New York Times*, February 26, 1999, A14.

64. Mark Jurkowitz, "Appetite Is Slight for a New Scandal," *Boston Globe*, February 22, 1999, A1.

65. Maria Seminerio, "Historians of the Internet Are Going to Call It the 'Monica Effect,'" *ZDNN*, August 19, 1998, reported at <www.zdnet.com>.

66. Felicity Barringer, "In Washington, Is There News after Scandal?" *New York Times*, February 15, 1999, C1. Also see Gay Jervey, "Chris Matthews Won't Shut Up," *Brill's Content*, September 1999, 78–83, 120–121.

67. Doris Graber and Brian White, "The Many Faces of News" (paper presented at the annual meeting of the American Political Science Association, Atlanta, September 2–5, 1999).

68. "Lewinsky is the Queen of Network Airtime," *Providence Journal*, December 23, 1998, A7. Also see Richard Berke, "Democrats' Gains Dispel Notion That the G.O.P. Benefits," *New York Times*, November 6, 1998, A22.

69. Richard Morin and Claudia Deane, "President's Popularity Hits New Highs," *Washington Post*, February 1, 1998, A1. Also see Regina Lawrence, Lance Bennett, and Valerie Hunt, "Making Sense of Monica: Media Politics and the Lewinsky Scandal" (paper presented at the annual meeting of the American Political Science Association, Atlanta, September 2–5, 1999).

70. Jane Fritsch, "Trust Me: A Media Guide," *New York Times*, February 1, 1998, WK5.

71. July 24–26, 1999 national survey by NBC and the *Wall Street Journal*, reported at <www.msnbc.com>, July 29, 1999. Also see Darrell M. West, *Patrick Kennedy: The Rise to Power* (Englewood Cliffs, N.J.: Prentice Hall, 2000).

72. Alan Pergament, "TV Coverage of JFK Jr. Tragedy Tries to Explain the Unexplainable," *Buffalo News*, July 19, 1999, 13A.

73. "Who Cares?" *Providence Journal*, July 22, 1999, A10.

74. Mark Jurkowitz, "Coverage Boosts Cable News Ratings," *Boston Globe*, July 21, 1999, A13.

75. Mark Jurkowitz, "Breaking News Brings Out Best and Worst of Local News Stations," *Boston Globe*, July 22, 1999, A8.

76. Ibid.

77. Alexandra Marks, "Media Lose Public's Respect in Coverage of the 'Big Story,'" *Christian Science Monitor*, July 22, 1999, 2.

78. Howard Kurtz, "The Times' Left-Hand Man," *Washington Post*, July 26, 1999, C1.

79. "No Autopsy Photos," *Cape Cod Times*, July 22, 1999, 1.

80. Gallup News Service, "Americans Saddened by JFK's Plane Crash; View Media Coverage Favorably, But as Excessive," July 27, 1999, reported at <www.gallup.com>, July 28, 1999.

81. Pew Research Center for the People & the Press, "Senate Trial: Little Viewership, Little Impact," press release, January 18, 1999, 1.

82. Jennifer Weiner, "Where Will All the Pundits Go?" *Providence Journal*, February 3, 1999, E1.

83. Katharine Q. Seelye, "Dole Urges Voters to 'Rise Up' against 'Liberal' Media," *New York Times*, October 26, 1996.

84. Pew Research Center for the People & the Press, "Democratic Congressional Changes Helped by Clinton Ratings," April 3, 1998 press release, 5. Also see Kenneth Dautrich and Thomas Hartley, *How the News Media Fail American Voters* (New York: Columbia University Press, 1999) 14, chapter 7; and Bruce Sanford, *Don't Shoot the Messenger: How Our Growing Hatred of the Media Threatens Free Speech for All of Us* (New York: Free Press, 1999).

85. Pew Research Center for the People & the Press, "Fewer Favor Media Scrutiny of Political Leaders," March 21, 1997 press release, 1.

86. Ibid., 8. The later study is cited by Dylan Loeb McClain in "Scandals Don't Much Harm an Already Bad Reputation," *New York Times*, October 19, 1998, C4.

87. Janny Scott, "In Scandal Coverage, Risky Era for News Business," *New York Times*, December 24, 1998, A14.

88. Michael Peltier, "Millions Log On to View Execution Photos," *Boston Globe*, October 26, 1999, A9.

CHAPTER 7

The Future of the Media

Over 150 years ago, the famed French observer Alexis de Tocqueville traveled to the United States for a firsthand look at the New World. Visiting hamlets all across the country, he compiled his impressions into what became the classic book entitled *Democracy in America*. In that volume, de Tocqueville emphasized how important newspapers were to the American experiment with popular rule, but how personally ambivalent he felt about the press of his day. Writing of newspapers in 1831 and 1832, he confided to his readers that "I do not feel toward freedom of the press that complete and instantaneous love which one accords to things by their nature supremely good."[1] In his view, newspapers were overly personalistic and nasty in their coverage. Their poor quality reflected the coarse and vulgar origins of leading journalists, he thought.

To make his point, de Tocqueville quoted at length from the very first political article he read upon arrival in America. The story from the *Vincenne's Gazette* bitterly condemned President Andrew Jackson, saying: "He governs by corruption, and his guilty maneuvers will turn to his shame and confusion. He has shown himself in the political arena as a gambler without shame or restraint."[2]

Yet despite the passions of American press coverage, de Tocqueville argued that newspapers were crucial to democratic governance. Speaking of the press, he noted the key to its importance. "Its eyes are never shut, and it lays bare the secret shifts of politics, forcing public figures in turn to appear before the tribunal of opinion," he wrote.[3] This type of public accountability, he argued, was vital to the long-term success of any political system, especially one based on popular sovereignty.

How, then, did he reconcile the "violence" of daily press coverage with its fundamental role in a democratic political system? The key, in his mind, to checking press excesses was in diversifying the number of outlets. Using language that speaks volumes to the contemporary period, de Tocqueville concluded, "the only way to neutralize the effect of newspapers is to multiply their numbers."[4]

In discussing the future role of the media in this country, I extend de Tocqueville's keen observation about the partisan press of the early nineteenth century to the contemporary period. Similar to the formative stage of the United States, our current era features a fragmented media with varying degrees of professionalism, marked by personalistic coverage, tabloid tendencies, and limited public respect for journalists. Early American newspapers lacked independent financial resources and were forced to depend on political party leaders and government officials for economic subsidies. This limited their objectivity and generated partisan coverage designed to curry favor with political patrons. The results were low subscription levels, lack of reader respect and an eighteenth-century media with very limited independent power. Citizens either read papers with which they already agreed or relied on more than one paper to obtain a balanced point of view, thereby constraining the overall power of the press.

The situation today is similar, but with a slightly different twist. Journalists no longer are dependent on political parties for their revenues, but media organizations face fierce economic competition unleashed by deregulation. The proliferation of news outlets and rise of new media such as talk radio, newsletters, and the Internet have broken the news gathering monopoly that elite journalists won during the twentieth century. Anyone can report the news today and with the aid of new technologies, the result has been a fragmentation of the media marketplace. Coverage is more heterogeneous and the varying degrees of professionalism among news reporters has weakened public esteem for the press and undermined the political clout of reporters. Through a variety of reasons which I discuss in this chapter, the media establishment has declined as a major political power. But because we have "multiplied their numbers," the excesses of individual outlets may not be problematic for the system as a whole. As de Tocqueville pointed out in the nineteenth century, there is safety in numbers.

THE DECLINING POWER OF THE MEDIA ESTABLISHMENT

For years, mass communications scholars argued vehemently against a "minimal effects" perspective which proposed that the media had only a limited capacity to alter people's views. Using work from the 1940s and 1950s, researchers documented that people were not ciphers who let journalists tell them what to think, but rather brought to newspapers, radio, and television preformed attitudes that allowed them to filter away information with which they disagreed. Only in narrowly prescribed circumstances, such as situations when there were no preexisting beliefs, could journalists have much of an impact on what people thought.[5]

Over the past three decades, though, this perspective has given way to a variety of new interpretations suggesting ways the media establishment are quite influential in shaping public opinion. Agenda-setting studies document that the media can affect the public's sense of priorities.[6] Other research details avenues by which the way a story is cast, or "spun" influences what readers and viewers take away.[7]

In conjunction with apparent demonstrations of media influence on events such as Vietnam, Watergate, and the Gary Hart sex scandal, the conventional wisdom suggested the reporters were dominant political players. Far from occupying minor political roles, journalists became recognized as a major political force. In this situation, campaigns were described as "mass media elections" and governing was said to be dominated by a media-based "spin cycle."[8] Through their high source credibility, control over political communications, and hold on the viewing audience, reporters were powerful.

However, as with many fields, the media industry has been rocked by the dramatic contemporary revolution that has taken place throughout contemporary culture. Many of the conditions over the past three decades that afforded the media great influence are now weakening. The unusually high source credibility of reporters has dissipated. The proliferation of media outlets has given viewers more sources of information. The television audience has splintered. Coverage is more heterogeneous, which means the public gets mixed messages from different information outlets. The clout of the establishment press has weakened in the face of new technologies, deregulation, and loss of control over news gathering. While journalists still have the ability to set the agenda through saturation coverage, their power to frame issues has weakened.

The overall result is that there is far less danger of media power today than at any point in American history. While the rise of the fragmented media may not exactly herald a return to the days of the minimal effects model, the future of the media looks strikingly like the model of FM radio — lots of outlets with no single station being dominant. Worries about an omnipotent and dominating press seem strangely out-of-date, given these new media realities.

THE EROSION OF MEDIA PROFESSIONALISM

There are many reasons behind the weakening power of the mainstream media from a decline in source credibility and rise of narrowcasting to heterogeneous coverage and smaller audiences as viewers take advantage of more media options. But one of the most important factors in the decline of the media establishment is the loss of industry professionalism. A press that emphasizes tabloid coverage and heterogeneous reporting from a large number of inexperienced journalists cannot exercise the same hold on the general public as an industry focusing on professional and homogeneous coverage dictated by a small number of elite outlets.

Political scientist Tim Cook recently made the point that the media have become a major institution, complete with fixed rules, professional norms, and powerful members. Political institutions, according to Cook, are characterized by three particular qualities: (1) stable procedures, routines, and assumptions about how to act socially and politically; (2) endurance over time and extension across organizations; and (3) exercising power over a societal and/or political sector.[9] Or to put it more simply, institutions have "stable, recurring, and valued patterns of behavior" that make them powerful political bodies.[10]

There is little doubt in this century that the mass media satisfy the criteria of a political institution. Certainly, during the time of the objective media, press organizations clearly fell within this category. With the professionalization of news routines and the creation of explicit guidelines about how to cover the news, reporters generally followed the dictates of organizational routines for news gathering. These professional norms and assumptions existed across news organizations and held up for a period of many decades. With the high source credibility of reporters at this time, media power over the political process was quite strong.

Even in the next stage, that of the interpretive media, professional norms were quite strong. Reflecting new intellectual currents and a changing sense of how news should be covered, the news industry shifted towards coverage that was more contextually based, interpretive in nature, and centered around news analysis and commentary. Both television and newspapers developed new modes of coverage such as ad watches that exemplified emerging professional thinking about appropriate styles of reporting.

During this period, there were a number of cases where the media demonstrated clout through character investigations of prominent officials and candidates. Gary Hart was the most prominent example, but he was not the only one. When John Tower was nominated for secretary of defense, press reporting of his drinking problem and womanizing effectively derailed his confirmation. Douglas Ginsburg was denied a seat on the U.S. Supreme Court after reports that he had smoked marijuana surfaced. Ethics investigations created problems for numerous public servants across the country, such as former House Speakers Jim Wright and Newt Gingrich, who were accused of profiteering on personal book deals.

Yet it is not as clear that media institutionalization and professionalism is holding up very well during the era of the fragmented press. Many of the professional norms, procedures, and routines that existed throughout the twentieth century are disappearing in the face of cutthroat industry competition, layoffs of experienced reporters, downsizing of media organizations, and restructurings that eliminate editors and fact checkers. The blurring of traditional lines between "news" and "entertainment" has further affected the way journalists do their jobs. As political scientists Michael Delli Carpini and Bruce Williams point out, the new media environment features a wide range of information providers, with varying degrees of professionalism.[11]

Indeed, one of the dominant public concerns about the news media today is loss of professionalism. Nearly half of newspaper editors believe that "the ability to publish information immediately online has led to an erosion of the standards of verification."[12] The public also worries that reporters are violating long-time industry rules on the verification of facts and use of anonymous sources. As discussed in Chapter 6, polls demonstrate that a clear majority of ordinary citizens believe the media today do not get their facts correct. Less than half approve of new media investigative tools, such as using hidden cameras and paying sources for news. Whereas 72 percent of Americans in 1985 described news organizations as "highly professional," in 1999 that figure had dropped to 52 percent, a stunning erosion of 20 percentage points.[13]

David Broder, the well-respected columnist for the *Washington Post*, put it most clearly in a 1998 speech at Harvard University discussing how journalism has changed over the past two decades. Commenting on the the rise of the Internet, talk radio, and cable television, and the way entertainment shows have affected news coverage, he said, "the more we structure our journalism to fit the conventions of sitcoms or TV dramas, the more we induce reporters to bark at each other like politicians; the more we take the meaning out of journalism and blur the definition of what a journalist is."[14]

With the dramatic increase in number and type of news outlets, adherence to industry norms and routines has weakened. There simply is no way for the industry to police reporters' behavior given the variety of organizations that present the news to the public. In the Internet age, for example, one person such as Matt Drudge with no training in the news industry can run a political news Web site from his apartment and attract hundreds of thousands of visitors a day.

One of the most concrete manifestations of this changing industry is illustrated by elite journalist dissatisfaction with the White House Correspondent's Dinner, long a major gathering for media bigwigs. In 1999, the rising number of entertainers and celebrities in attendance in recent years, such as Ellen DeGeneres, Paula Jones, and Larry Flynt, led the *New York Times* to boycott the dinner on grounds that the event no longer served a legitimate industry function.

A fragmented media containing different degrees of professionalism makes it difficult for scholars to generalize about the industry based on single news outlets. It has been common in academic studies to choose one outlet, such as the *New York Times* or ABC News, and argue that outlet was representative of all media organizations. During the period of homogeneous coverage, that was a fairly reasonable assumption; it really didn't matter which outlet got covered. Now, however, the rise of heterogeneous reporting by different media makes it very risky to generalize from any single outlet. The media field has become like the proverbial story of the blind men touching the elephant. Individuals have different reactions depending on whether they are focusing on the trunk, the legs, the hide, or the tail. The same has become true of the media. Depending on whether observers examine NBC, Rush Limbaugh, the *New York Times*, or the Drudge Report, different conclusions will be reached about the content and tone of news coverage.

PROTECTION IN LARGE NUMBERS

Despite clear evidence of weakening media professionalism and obvious tabloidism from single outlets, the current media era is less dangerous politically than commonly believed. The reasoning picks up on the logic offered long ago by de Tocqueville. The excesses of single media units can be limited by the large number of news providers, diversity of coverage in the system as a whole, loss of audience share by major news organizations, and the overall reduction in media power.

The dominant critiques of the mass media in recent years have emphasized the problems of corporate control over the news as well as the superficiality, tabloidization, cynicism, and biases of news outlets. According to a number of critics, news coverage has suffered because powerful conglomerates own news organizations and either subtly or overtly shape the way political events get reported. Scholars such as C. Wright Mills, Noam Chomsky, and Ben Bagdikian and more recent writers such as Mark Hertsgaard and Robert McChesney have complained that the combination of economic oligopolies and intellectual vapidity has ruined the educational potential of the media.[15] Reporters are not objective in how they cover events due to their dependence on corporations for their livelihoods and the interlocking nature of many media empires.

While there is no doubt that corporate control continues to be a serious matter and that corporations are interested in taming the unfettered competition of the Internet, the competitiveness of the current system, the large number of news outlets, the heterogeneity of coverage, and the absence of barriers to entry into the world of news gathering has weakened corporate power over the news. In an era of structural fragmentation and proliferation of new outlets, there is protection in large numbers. As de Tocqueville argued in the nineteenth century, if there are many news providers, no one of whom dominates the media system, there is security against the power of the few. Certainly there is much greater diversity within news organizations today. The shift from three to seven television networks, the declining audience share of ABC, CBS, and NBC, and the rise of cable, talk radio, and the Internet has broken the monopoly of the media elite and undermined the corporate power of the conglomerates that employ them.

In addition, media critics have accused reporters of being "out of order" and "undermin[ing] American democracy." According to Thomas Patterson, one of the most insightful contemporary critics of media political coverage, serious problems plague the mass media. Writing in his best-selling 1993 book, *Out of Order: An Incisive and Boldly Original Critique of the News Media's Domination of America's Political Process*, Patterson argues that "journalists are the problem" in contemporary politics, not politicians.[16] Reporters are unduly cynical about public officials, always assuming that they are lying or trying to hide something.

In an analysis of coverage by *Time* and *Newsweek* of major party nominees from 1960 to 1992, Patterson finds a dramatic increase in unfavorable candidate coverage. For example, 75 percent of press reporting on Kennedy and Nixon in 1960 was positive. This percentage dropped steadily over the past three decades to the point where in 1992, only 40 percent of the coverage about Clinton and Bush was positive.[17]

Not only is much of the coverage unduly negative, asserts Patterson, but campaign coverage is based on "the horse race" between candidates, that is, who is ahead and what strategies are they using. This type of coverage drives out substantive coverage of the issues argues Patterson, who concludes his analysis by saying "the United States cannot have a sensible campaign as long as it is built around the news media."[18] Coming from an author whose previous book was entitled *The Mass Media Election*, this conclusion is damning indeed.

James Fallows is no more positive about the contemporary media. In his 1996 book, *Breaking the News: How the Media Undermine American Democracy*, he claims the press has been "irresponsible with its power."[19] Reporters badger candidates and public officials in an effort to get answers. Politics is considered a cynical game wherein "ambitious politicians struggle for dominance, rather than a structure in which citizens can deal with worrisome collective problems."[20]

Journalists have financial conflicts of interest that cloud their reporting. He cites celebrity journalists, such as George Will, Cokie Roberts, and Sam Donaldson, who earn millions of dollars from salaries, lectures, and books. Contemporary journalists are about as far removed from de Tocqueville's "coarse" and "vulgar" working class as possible. Not only are leading reporters no longer poor and uneducated, a number of them have become celebrities in their own right.

According to Fallows, financial conflicts are common in the journalism fraternity. One example he cites in his book is Donaldson, who did an ABC *Prime Time Live* segment attacking farm subsidies, but never disclosed that he himself had claimed $97,000 in sheep and mohair subsidies from his New Mexico ranches. Fallows also criticizes Will for not revealing during television discussions of and columns on tariffs for Japanese luxury cars that his wife, Mari Maseng was a paid lobbyist for the Japanese Automobile Manufacturers Association.[21]

These various criticisms of the mass media resonate with the general public. Two-thirds of Americans see political bias in news coverage.[22] Respect for journalists is down, and viewers are deserting mainstream media outlets for a variety of new media outlets, such as talk radio and the Internet. This, of course, intensifies the competitive pressures on elite outlets, and makes it more likely that reporting is done by inexperienced reporters and filled with errors, rumor, and innuendo. It appears that the industry is trapped in a vicious cycle of hyper-competition from which it cannot escape.

While there is little question that individual media outlets during the age of fragmentation have genuine problems with their coverage, this critique misses key elements of this media era. For one, this line of reasoning devotes far too much attention to dimensions of media evaluation centering on objectivity in coverage. Little attention is devoted to the equally important criterion of diversity of views conveyed by the media overall.

As noted in Chapter 6, with the proliferation of media outlets, the number of news options is up dramatically over thirty years ago. The current era is closer to satisfying de Tocqueville's key condition for ameliorating press excesses than at any point since the mid-nineteenth century. At no other time has the American public had as many reading and viewing options for public affairs information, business news, and entertainment. The range of options is simply staggering.

It is not just the increase in number of outlets that protects against shocking excesses from any particular news outlet. With this transformation has come a marked change in favor of more heterogeneous coverage of the news. No longer do leading outlets cover public affairs in exactly the same manner and from the same perspective. Homogeneous coverage is out; variety and differentiation is in.

Regardless of whether one's perspective is liberal, conservative, libertarian, fundamentalist, feminist, gay, Nazi, communist, environmental, or vegetarian, there are newsletters, periodicals, radio stations, Internet sites, and public access cable channels which serve that point of view. As de Tocqueville argued many years ago in regard to the partisan press, the best protection for democracy from a powerful and homogeneous press is competition. The more media outlets there are, the greater diversity of views. No era of American history has seen the diversity of news perspectives common today.

In this highly competitive marketplace, media outlets are finding particular niches in the market and narrowcasting to that area. For those who want in-depth coverage of public affairs, there are two C-SPAN channels which cover live proceedings of the House and Senate along with a variety of public policy conferences and speeches from around the country. All-news channels CNN and MSNBC devote twenty-four hours a day to news broadcasting. CNBC covers news related to finance and business.

Nearly every major newspaper has developed a Web site providing a wealth of information, more than even the most curious citizen generally desires. Using powerful search engines, people can search these sites as well as millions of others for information on a pet cause. If it is the militia movement one wants information on, it is easy to find sites devoted to that subject. Or if one's interests lay more toward libertarianism, there are Web sites, newsletters, and magazines devoted to that cause.

Even talk radio and Web chatrooms, which long have been disparaged by academics, contribute to the creation of "public spheres," arenas where social and political discussion take place. Such domains have worried critics who complain about the frenzy or ideological fervor with which public discussions take place. Yet these very qualities are essential and should not be unexpected as society grapples with emotional subjects such as race, school violence, date rape, and health care. Even with their obvious political slants, hosts such as Rush Limbaugh, Don Imus, Howard Stern, Jay Leno, and David Letterman contribute more to democratic discourse than they have been given credit for.

Time-worn critiques about the media not covering issues or providing inadequate substantive information are no longer true. In fact, these conclusions are an artifact of studying only a narrow band of media outlets, such as a handful of news magazines or television network news shows, and ignoring the wide range of viewing and print choices available. By practically any criteria of evaluation, there are more than enough in-depth information sources in the contemporary period to satisfy even the most demanding news consumers around the country.

Indeed, when one looks back at the "good old days" of the objective press, the programming from that time was not as healthy for democracy as people today want to remember. The existence of a few dominant outlets and courteous competition between these outlets created a homogeneity in coverage that suppressed diversity of opinions. Important background information on powerful politicians was not made public.

The need of the media industry to appeal to a broad mainstream audience led to bland and noncontroversial programming. Such coverage reflected the overwhelmingly white and male perspective of industry leaders. There was no diversity of political opinion, little choice for consumers, and hardly any in-depth coverage. Nostalgia for the days before the tabloid press ignores how limited the coverage of that day actually was. We should abandon treating the era of the objective press as the golden age of the mass media in terms of the quality of the coverage.

RISKS FACING THE FRAGMENTED MEDIA

Diversity protects the nation as a whole, but do the tabloid tendencies and loss of public respect for specific units create any special risks for contemporary journalists? In my view, yes. The clear backlash that has developed against the media over the past decade has raised several threats to the press as we currently know it.

Tougher Libel Laws and Jury Awards

Rising public concern about how journalists handle their jobs exposes the industry to tougher rulings in slander and libel cases, and through proposed laws restricting overly aggressive paparazzi. For example, in 1998, the House Judiciary Committee heard testimony from prominent actors Michael J. Fox and Paul Reiser, among others, complaining of tabloid photographers "sneaking into hospitals to photograph their newborn children — or spitting in the actors' faces to provoke a dramatic photo." In a response to these actions, Fox explained, "I expect to have my photograph taken . . . [But] I strongly disagree with those who would argue that some sort of Faustian bargain has been struck whereby public figures are fair game, any time, any place, including within the confines of their own homes."[23]

In speaking out, Fox supported a proposed bill that would make it a federal crime "for a photographer to threaten or cause bodily injury in the pursuit of photographs or recordings." The legislation, sponsored by Representative Elton Gallegly (R-Calif.), sought to ensure that celebrities had "a reasonable expectation of privacy" and would not have to fear death or bodily injury as a result of being chased by paparazzi.

Others have sued media outlets on grounds of defamation of character and journalistic misrepresentation. In 1998, the owner of a Maine trucking company sued NBC's *Dateline* show claiming that a negative show about drivers in the industry falsifying logbooks, getting little sleep, and violating federal rules limiting driving time had destroyed his reputation and his business. The segment broadcast on April 19, 1995, entitled "Keep on Truckin'" proclaimed, "what you're about to hear about the drivers of many of these trucks won't make you feel any safer."[24] Not only did the owner claim the segment had ruined his trucking business, he said it had destroyed his faith in the news media. "I don't believe what I see anymore," he announced. Shortly thereafter, a jury found NBC guilty of negligence and awarded $525,000 in damages.

Also that year, the *Cincinnati Enquirer* paid Chiquita $10 million and made an extraordinary front-page apology in order to avoid being sued for articles critical of the company's business practices. In an investigative article published May 3, the newspaper made the accusations, which had been based on two thousand internal voice mail messages, but retracted the charges on June 28 after concluding the messages had been stolen by one of the reporters who wrote the story.[25]

Representative Corrine Brown (D-Fla.) went even further in her complaints about the media. After two reporters made her cry during questioning for an investigative piece on her personal finances, she filed criminal charges with U.S. Capitol Police and attempted to strip their press credentials for covering Congress.[26]

Although understandable from the point of view of those who feel aggrieved at overly intrusive press coverage, toughening of rules against the media is dangerous to the long-term health of the country. It should be resisted due to its importance for the ability of the press to hold leaders accountable. Without a chance to investigate public officials or check into people's backgrounds, information that is relevant for campaigns and governance would be lost, to the detriment of democratic accountability.

The excesses of individual media outlets are not grounds for trampling on the constitutional doctrine of freedom of the press. Journalists are the central overseers of public officials. Reporters who ask probing questions and investigate behind-the-scenes conduct is crucial enough to democratic governance to warrant the tolerance of scurrilous press coverage. The worst outcome for the country would be punishing the media collectively for the defamatory and obnoxious excesses of individual reporters, which can be dealt with by existing statutes.

Public Boycotts

Another risk for the media that has emerged in contemporary times centers on the increasing tendency of media outlets to focus on niche news markets. This development in the fragmented media exposes specific outlets to much greater danger from discontented groups within society. The shift from broadcasting to narrowcasting is fundamental to the way the industry is changing. As long as the press consisted of a few large organizations, public boycotts had almost no hope of ever influencing news outlets. The breadth of the audience and the wide scope of broadcasting effectively insulated news outlets from unhappy groups in society.

However, now there are many more news organizations and each has a much smaller audience. The networks have lost audience share and broadcasters have responded by searching for smaller niches that they can dominate (like congressional deliberations for C-SPAN, sports for ESPN, and weather for the Weather Channel).

This puts consumers in a much stronger position either to threaten or actually harm a single outlet. Small media organizations are quite dependent on advertisers and audience. The risk of political blackmail and/or extortion from unhappy viewers is significant today. It is little surprise that public boycotts have arisen in response to these industry changes.

For example, in 1997, the Southern Baptist Convention gained prominence when it called for a boycott of the Walt Disney Company, including its "movie studios, cable television channels, book publishers, trade magazines, newspapers, television and radio stations and the ABC network." The reason for the boycott was that Disney was offering health benefits to homosexual partners of its employees and that its ABC television show *Ellen* featured a lesbian who came out of the closet. With 15.7 million members, Baptists felt it was time to take a stand. "We believe homosexuality is a choice and that it's a bad choice," explained Tom Elliff, president of the Southern Baptist Convention. His associate, Richard Land, president of the convention's morals and ethics panel, added, "When Disney crosses to the other side of the street, there's a sense of betrayal and outrage. You can't walk the family side of the street and the gay side of the street in the Magic Kingdom at the same time."[27]

Two years later, an Arab-American organization known as American Muslims for Jerusalem called for a boycott of Disney because an Israel exhibit at the Epcot Center "depicts east Jerusalem as part of Israel, designates Jerusalem as the nation's capital and marginalizes the roles of Muslims and Christians in Jerusalem."[28]

These boycotts were not isolated episodes. The tactic has become increasingly popular among different groups in society. In Hartford, Connecticut, WZMX-FM radio host Sebastian was threatened with a boycott by an organization called PLEASE (Promoting Legislation & Education About Self-Esteem) after he made comments over the air complaining about people "who are heavy." Reportedly, he claimed he would not hire a fat person for a job and that fat people "hate themselves and hate life."

PLEASE, which devotes itself to increasing awareness of discrimination based on physical appearance, quickly gathered six hundred signatures from people who said they would stop listening to the show unless there were an on-air apology. It also notified twenty advertisers on the show about the group's boycott. For his part, Sebastian argued that the group simply was trying to intimidate him. "I think [they are] trying to bully me into making an apology for a remark I made that is my opinion. I'm not breaking any laws. . . . I entertain people with a caustic, abrasive, sarcastic look on life," he said.[29]

The Howard Stern radio show faced the threat of a boycott from the National Hispanic Media Coalition after Stern made remarks interpreted as anti-Latino following the murder of Tejano singer Selena. According to the group, Stern mocked her music a few days after she was shot to death. Referring to her singing, Stern said, "This music does absolutely nothing for me. Alvin and the Chipmunks have more soul. Spanish people have the worst taste in music."

Coalition chairman Alex Nogales explained that a boycott of advertisers was necessary because Stern "just went after every Latino in the nation. We have to understand our incredible power and we need to encourage ourselves to go do it." Within a few months, companies that withdrew their ads from Stern's nationally syndicated radio show included Gatorade, Sears, Quaker Oats, Pizza Hut, Anheuser Busch, Men's Warehouse, and Mitsubishi.[30]

With the fragmentation of the media marketplace, a decline in professionalism, and increasing public discontent, news outlets are more susceptible to boycotts than at any point in American history. The niche emphasis of many broadcasters opens them up to use of this tactic in a way that would not have been very likely when the industry focused on broadcasting to large segments of society. With narrowcasting very much in vogue due to structural changes in the industry, boycotts become a much more viable strategy for unhappy groups.

The Decline of a Shared American Culture

Perhaps the most fundamental risk for society as a whole is the decline of shared understandings of social and political events. One of the advantages of homogeneous coverage was that it provided a common framework for receiving and interpreting the news. Regardless of whether you watched ABC, CBS, or NBC, or read the *New York Times* or *Washington Post*, the news that you heard or read would have been similar. Professional norms governing what was newsworthy dictated a more or less common news product at the end of the day.

The same was true in terms of popular culture. Shows featuring Andy Griffith and Lucille Ball promoted a cultural homogeneity that provided common points of reference for many Americans. Even though many of these perspectives were mythical in nature, the shared sense of understanding integrated society and gave viewers a similar view of the world.

In contrast, the expansion in the number and diversity of media outlets today means there are many more subcultures, each of which views politics and society from its own particular frames of reference. A gay newsletter or cable channel obviously covers *Ellen* differently from a fundamentalist outlet. In this era of extensive media fragmentation and the Internet, it is much more difficult to forge common bonds across the country. In fact, studies have found that heavy Internet users spend less time with other people.[31]

Already, there are signs of a deep racial gap in public access to the new technologies of the Information Age. According to one study of Americans' access to computers, by a 44 to 29 percent margin, whites are more likely to own a home computer than blacks. Among college students, this gap is even wider; whereas 73 percent of white high school and college students have access to a home computer, just 32 percent of black students do. Another study published in 1999 found that "black and Hispanic families [were] less than half as likely as whites to explore the Net."[32]

The danger in the long-run from media fragmentation is that the ties that bind Americans together will weaken, groups will have less in common, and social conflict will become more significant. Signs of this have appeared, as evidenced by the emergence of a gender gap in political attitudes, racial gaps in how whites and nonwhites feel about discrimination and the legal justice system, and a generation gap in opinions about Social Security.

For example, research has documented a "major racial divide" between whites and blacks in views of police discrimination. Blacks are much more likely to believe that the police are unfair in enforcing laws, not courteous in dealing with motorists they detain, and biased in who they pull over for traffic violations. Whereas 61 percent of whites give the police positive marks for treating "all drivers the same regardless of race, sex or age," 72 percent of blacks rate the police negatively. In addition, 56 percent of blacks complained that the police "treat minorities worse than others."[33]

The same gulf is present in views about racism in society. While 38 percent of whites think it is a "big problem," 54 percent of blacks do. When asked if minorities have the same opportunities to live a middle-class life as whites, 60 percent of whites believed blacks had the same chance, compared to 35 percent of blacks who felt that way.[34]

A graphic illustration of the way race influences how people think surfaced recently in response to a false Internet rumor that Congress planned to repeal the Voting Rights Act in 2007 and take away blacks' right to vote. Spread persistently via new electronic media, this allegation generated so many worried calls to congressional offices that the Congressional Black Caucus was forced publicly to deny the rumor.[35]

One study by Russell Neuman, *The Future of the Mass Audience* published in 1991, predicted that this "loss of shared culture" would not be very likely because even though the number of television options was increasing, people still were sticking with a relatively small number of broadcasters. According to that author, "the movement toward fragmentation and specialization will be modest and . . . the national media and common political culture will remain robust."[36] Among other things, the sheer costs of producing and promoting new shows placed constraints on "special-interest, small-audience programming."

While valid in the context of the time period studied, the shift towards narrowcasting and the rise of low-cost alternatives on cable television and the World Wide Web refute the predictions of that research. If anything, the trend since 1991 has been in the direction of greater fragmentation and reliance on broadcasting niches, even by national television networks. The danger, then, to our national political culture is much greater than was envisioned at the beginning of the 1990s.

The more disparate the system of political communications becomes, the more difficult it will be for Americans to share a common culture. One of the qualities that has marked most of American history since 1850 has been a media that provided similar points of reference and a shared sense of understanding about political events. With the development of gender, racial, and generational gaps in our political communications, those cultural underpinnings are beginning to diverge substantially.

In the long run, electronic Balkanization of information poses many risks for society. It creates the potential for people of different backgrounds to misunderstand one another. It makes it more difficult for political leaders to build broad-based coalitions. It weakens the cultural ties that have bound America together for more than two hundred years. In short, this situation raises a host of serious problems for our community life.

FUTURE SCENARIOS FOR THE MASS MEDIA

Given the dramatic changes that have swept the media industry in recent years, it is foolhardy to attempt to predict what lies ahead in the future. The brisk pace of technological, economic, and organizational developments guarantees that little will remain stagnant in the decades ahead. However, within the rubric of many possible changes, three scenarios stand out as major possibilities.

Cutthroat Competition

One scenario basically projects current trends and presumes a continuation of media fragmentation and cutthroat competition into the foreseeable future. In this situation, the Internet will continue to proliferate. We will move from one hundred to five hundred cable channels. Desktop publishing will turn everyone who wants to be a newspaper editor into one. Niche news coverage and narrowcasting will be the norms under this system.

If this scenario is accurate, all of the risks discussed earlier will become even more problematic. It will be virtually impossible for the media industry to police its worst excesses. The media will be like an octopus without a head to guide the actions of the tentacles. The threat of crackdowns on freedom of the press will become more intense. Public boycotts will become more prevalent. And American culture will dissipate into a bewildering array of electronic subcultures, each of which has little in common with the other.

Although many people within the industry and among the viewing public will be dissatisfied with the media system that arises from this scenario, there will be little any professional organization can do. Tabloid coverage, inaccurate reporting, and reliance on unpopular investigative tactics such as hidden microphones and paying sources for stories will persist and media organizations will be unable to police their own industry.

The irony is that even as media choices surpass any other point in American history, public satisfaction with the news product will hit all-time lows. Similar to the ecological metaphor of the "tragedy of the commons," the media industry will suffer from the rampant competition among individual news outlets. Such unfettered competition is a surefire recipe for individual reporters to create major problems, and potentially bring down the entire media industry.

Industry Reconcentration

A second scenario assumes that intense competition will eventually prove unstable and give way to a reconcentration within the industry around a few media outlets. For example, seeing possibilities for cross-industry economies, Disney may continue to pursue its media empire and extend its ownership beyond ABC to major newspapers, magazines, radio stations, and Internet sites.

Using the same reasoning, Bill Gates, Paul Allen, and Microsoft may invest their billions in building not just MSNBC, the jointly operated channel with NBC, but other promising media properties, such as major newspapers, a wire service, a television network, or a major cable system. Indeed, the alliance between NBC News and Microsoft has produced a creative cross-marketing strategy based on promoting parallel stories on the *Today Show*, *NBC Nightly News*, *Dateline*, MSNBC cable shows, and the MSNBC Internet site. Once third in the network news audience rankings, NBC now has surged to the top in part due to its ability to link together different media vehicles around common themes.

Ted Turner may use the money he made from CNN and Time Warner to buy new communications outlets or merge with another media conglomerate. Rupert Murdoch may invest the billions he made from the tabloids in new information properties. He already has made a major move toward mainstream respectability by purchasing the Los Angeles Dodgers, one of the premium name brands of major league baseball.

The vast explosion of wealth created in the 1980s and 1990s makes it possible for the current system of cutthroat competition to prove ruinous, eventually leading to a small number of "megamedia" organizations that will dominate political communications.[37] In place of a continuation of the current structure, this scenario anticipates a few large oligarchies that will dominate the media marketplace.

Here, the danger will come not so much from intense competition and niche markets, but a stifling of competition and concentration of excessive communications influence in the hands of a few people. In the same way that people around the turn of the twentieth century feared John D. Rockefeller and Standard Oil, a media industry dominated by Disney, Microsoft, Turner, and Murdoch would prove unsettling to many Americans. This large concentration of media power and the potential to abuse the public trust will turn citizens long suspicious of concentrated economic power into open skeptics.

Similar to the turn of the twentieth century Progressive movement, this scenario would breed grassroots citizens' protests. There would be a rise of organizations such as Ralph Nader's consumer safety group and Ross Perot's Reform Party, each dedicated to exposing the corrupting power of media monoliths.

Yet it would take extraordinary citizen interest in the political process for these efforts to prove very effective. It could be years before public discontent reached the boiling point where regulatory curbs could be put in place. After all, it was several decades before Progressive Era reformers achieved significant changes. And even then, World War I wiped out many of these gains.

European-Style Partisan Press

A third possibility is that the American media will follow its European counterparts and evolve into a modern version of the partisan press. For example, in England, the *Times* is widely seen as the Tory paper, the *Guardian* as the Labour paper, the *Independent* as the paper of the political center, and various other tabloids representing an array of antiestablishment perspectives. There is a clear delineation of ties between each paper and partisan organizations. Much in the same way that the first epoch of the American press exhibited close links between newspapers and political parties, these British outlets reflect a definite partisan point of view.

In the same manner, the presses of France, Germany, and Italy exhibit strong ties between the media and party organizations. There are papers reflecting communist, socialist, Catholic, business, and trade union perspectives. Everyone who reads a particular paper understands where that newspaper is coming from and judges its contents accordingly. It is a media model that maximizes diversity of opinions in the system as a whole and features substantive material in its coverage.

The fact that each outlet is biased according to its distinctive ideological perspective does not bother readers because they are free to select from a wide range of choices of coverage they voluntarily accept. It is no coincidence in these political systems that no one worries about bias or media power. Reporters are not seen as very influential politically, and bias in news coverage is readily advertised in advance to anyone who picks up the paper. Although different from the last 150 years of the American press, it is a system that works quite well from the standpoint of informing the public and representing diversity of opinions in the media system.

Applying this reasoning to the contemporary period, this scenario predicts that between the rise of media commentary and interpretation on the one hand, and the need to find news niches on the other, American media will evolve into a partisan media system with a clear political perspective represented in different media outlets. The fact that partisanship has declined among the general public will not necessarily stop the rise of this system because Americans increasingly are sorting themselves out into distinctive niches based on race, gender, lifestyle, geography, and political views, among other things.

Under this reconcentration scenario, Murdoch properties would reflect their owner's conservative policies. Turner enterprises would mirror the liberalism of that man. Following government efforts to break up Microsoft for anticompetitive practices, Gates's media outlets would be solidly libertarian in nature and opposed to governmental intervention in the economy. Disney would represent its pragmatic vision of commercialism and support politicians favorable to its point of view.

Alternatively, if fragmentation became the rule of the day over reconcentration, there would be hundreds of partisan outlets, each reflecting a small niche in the ideological marketplace. No one would dominate, a thousand flowers would bloom, and there would be lots of diverse views represented in the political system, from Nazis and Ku Klux Klanners to environmentalists, feminists, and Promise Keepers.

If this scenario occurs, it would give America a media system for the twenty-first century that resembled that of the early nineteenth century. There would be many different political perspectives represented through various media organizations. Each outlet would cover politics from its particular partisan standpoint. The political danger to ordinary citizens would be minimized because readers and viewers would know what they were getting when they picked up a paper or tuned in a television station. It would be a system that would satisfy the criteria for an effective press laid down by de Tocqueville more than 150 years ago.

CONCLUSION: THE CRUCIAL MEDIA ROLE IN DEMOCRACY

In whatever manner media outlets change over the next few years, it is clear that the way in which they are organized will have consequences for how the political system functions and how the public sees the press. Because of their responsibility for communications, no matter how reporters handle their jobs, they structure the nature of political competition. They influence how issues are framed, conflicts are resolved, and citizens evaluate the political process.

For each historical epoch studied in this book, the media have helped shape political debates, often with important consequences. From the partisan and commercial presses of the nineteenth century to the objective, interpretive, and fragmented presses of the twentieth century, journalists have played an important role in our political system.

The shift from a weak press in the nineteenth century to a powerful press in the twentieth century represented a dramatic success for the American media. The style of coverage that emerged during that period gave journalists tremendous credibility and political clout, making the media's fall from grace over the past decade all the more compelling. With the dizzying array of media outlets and the range of professionalism that has emerged, journalists increasingly are offending both government officials and the general public.

This situation puts our political system at some risk. Given current media excesses and unsatisfactory coverage, it is tempting to crack down on the press, take away certain freedoms, and organize boycotts against coverage we do not like. However, we should resist such urges. Democracy cannot survive without a free press — that has been the reality throughout American history.

In the long run, nothing is more fundamental to the operation of the political system than the media. In an era when people pay little attention to politics, citizens need journalists for basic information about government and cues about how their leaders are performing. Reporters continue to provide early warning signals for ordinary voters. Our challenge is to find a way to harness new media realities to our common good.

NOTES

1. Alexis de Tocqueville, *Democracy in America*, edited by J. P. Mayer and Max Lerner (New York: Harper & Row, 1966), 166.
2. Ibid., 168.
3. Ibid., 171.
4. Ibid., 170.
5. Joseph Klapper, *The Effects of Mass Communications* (New York: Free Press, 1960).
6. Shanto Iyengar and Donald Kinder, *News That Matters* (Chicago: University of Chicago Press, 1987); and Maxwell McCombs and Donald Shaw, "The Agenda-Setting Function of Mass Media," *Public Opinion Quarterly* 36 (1972): 176–187.
7. Shanto Iyengar, *Is Anyone Responsible?* (Chicago: University of Chicago Press, 1991).
8. Thomas Patterson, *The Mass Media Election* (New York: Praeger, 1980); and Howard Kurtz, *Spin Cycle* (New York: Free Press, 1998).
9. Timothy Cook, *Governing with the News* (Chicago: University of Chicago Press, 1998), chapter 4.
10. Ibid., 66.
11. Michael X. Delli Carpini and Bruce Williams, "Let Us Entertain You: The Politics of the New Media Environment," in *Mediated Politics*, Lance Bennett and Robert Entman, eds. (New York: Cambridge University Press forthcoming, 2000).
12. Janet Kornblum, "Newspaper Editors See Standards Slip Online," *USA Today*, May 16, 2000, 3D.
13. Pew Research Center for the People & the Press, "Big Doubts about News Media's Values," press release, February 25, 1999, 8. Also see Pew Research Center, "Striking the Balance: Audience Interests, Business Pressures and Journalists' Values," 1999; and Bill Kovach and Tom Rosenstiel, *Warp Speed* (New York: Century Foundation Press, 1999). Similar complaints about outlandish behavior have arisen in regard to entertainment television, as demonstrated by Bill Carter in "Cable Television Ups the Ante on the Outrageous," *New York Times*, March 22, 1999, C1.
14. David Broder, "Theodore H. White Lecture," Joan Shorenstein Center for Press, Politics, and Public Policy, Harvard University, November 12, 1998, 15. Also see Jeffrey Scheuer, *The Sound Bite Society* (New York: Four Walls Eight Windows, 1999).
15. C. Wright Mills, *The Power Elite* (New York: Oxford University Press, 1956); Ben Bagdikian, *The Media Monopoly*, 2nd ed. (Boston: Beacon Press, 1987); Noam Chomsky, *Media Control* (New York: Seven Stories Press, 1997); Mark Hertsgaard, *On Bended Knee: The Press and the Reagan Presidency* (New York: Farrar, Strauss, & Giroux, 1988); and Robert McChesney, *Rich Media, Poor Democracy* (Urbana: University of Illinois Press, 1999).
16. Thomas Patterson, *Out of Order* (New York: Vintage Books, 1993), 16.
17. Ibid., 20.
18. Ibid., 25.
19. James Fallows, *Breaking the News: How the Media Undermine American Democracy* (New York: Pantheon Books, 1996), 9.
20. Ibid., 31.
21. Ibid., 37–38.
22. "Traditional News Sources Lose Ground," *Providence Journal*, February 6, 2000, A32.
23. "Actors Fox, Reiser Back Bills Restricting Aggressive Paparazzi," *Providence Journal*, May 22, 1998, A11. Also see Bruce Sanford, *Don't Shoot the Messenger: How Our Growing Hatred of the Media Threatens Free Speech for All of Us* (New York: Free Press, 1999).
24. Brian MacQuarrie, "Maine Truckers Take NBC Show to Court, Alleging Defamation," *Boston Globe*, June 18, 1998, B1. NBC also was criticized for eliminating references to nuclear waste from its *Atomic Train* miniseries because its parent company General Electric owns a nuclear power division. See

Lawrie Mifflin, "NBC Edits 'Nuclear Waste' from Mini-Series, and Outside Pressure Is Charged," *New York Times*, May 14, 1999, A19.

25. Laurence Zuckerman, "Paper Forced to Apologize for Articles about Chiquita," *New York Times*, June 29, 1998, A10. Also see Douglas Frantz, "For a Reporter and a Source, Echoes of a Broken Promise," *New York Times*, April 11, 1999, 3.

26. Jim VandeHei, "Rep. Brown Tries to Punish Reporters," *Roll Call*, July 2, 1998.

27. Allen Myerson, "Southern Baptist Convention Calls for Boycott of Disney," *New York Times*, June 19, 1997, A18.

28. "Arab Groups Calling for Disney Boycott," *Providence Journal*, September 22, 1999, A8.

29. Bill Keveney, "Group Boycotts WZMX over Sebastian Remarks," *Hartford Courant*, October 23, 1996, E2.

30. Armando Villafranca, "Radio Show, Stern Lose More Sponsors; Boycott Growing over Selena Remarks," *Houston Chronicle*, May 20, 1995, A30.

31. John Markoff, "A New Lonelier Crowd Emerges in Internet Study," *New York Times*, February 16, 2000, A1.

32. Amy Harmon, "Racial Divide Found on Information Highway," *New York Times*, April 17, 1998, A1. The 1999 study is reported in David Sanger, "Big Racial Disparity Persists among Users of the Internet," *New York Times*, July 9, 1999, A12.

33. *Star-Ledger*/Eagleton Poll Press Release, "New Jerseyans' Views of the State Police: A Contrast in Black and White," May 17, 1998, 1–3. Also see James Sterngold, "A Racial Divide Widens on Network TV," *New York Times*, December 29, 1998, A1.

34. Elliot Krieger, "'We Don't Know Each Other Very Well': Racial Groups Vary Widely in Views of R.I., Poll Shows," *Providence Journal*, June 21, 1998, A1.

35. "Internet Rumor Spreads Alarm among Blacks," A4.

36. Russell Neuman, *The Future of the Mass Audience* (New York: Cambridge University Press, 1991).

37. Dean Alger, *MegaMedia: How Giant Corporations Dominate Mass Media, Distort Competition, and Endanger Democracy* (Lanham, Md.: Rowman & Littlefield, 1998). Also see Richard Siklos, "Dot.Com or Bust," *Business Week*, September 13, 1999, 78–82; and Robert McChesney, *Rich Media, Poor Democracy* (Urbana: University of Illinois Press, 1999).

Appendix

FIGURE A.1 ▪ **Percentage of Population Subscribing to a Daily Newspaper**

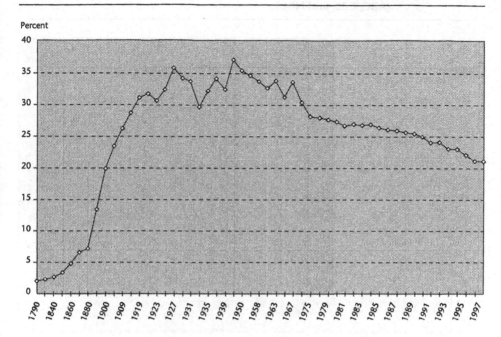

FIGURE A.2 ▪ **Percentage of Households Having Various Media**

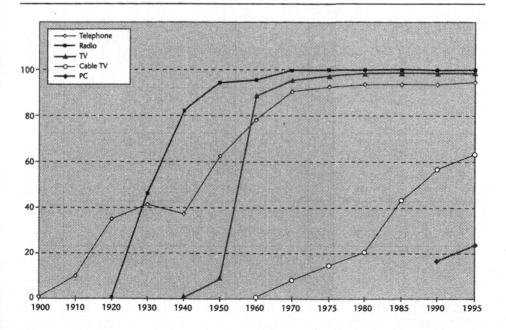

FIGURE A.3 ▪ Percentage Watching "Big Three" Television Networks (ABC, CBS, and NBC)

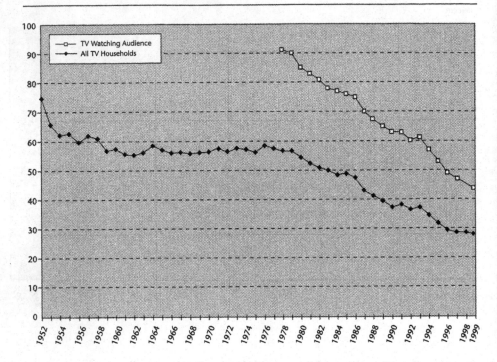

Index

Printed in the United States
By Bookmasters